COSMOLOGY AND ASTROPHYSICS

THOMAS GOLD. Photograph by Herman J. Eckelmann, Ithaca, New York.

COSMOLOGY AND ASTROPHYSICS

ESSAYS IN HONOR OF THOMAS GOLD

Edited by
YERVANT TERZIAN
and
ELIZABETH M. BILSON

Cornell University Press
ITHACA AND LONDON

6666-9236

First published 1982 by Cornell University Press.
Published in the United Kingdom by Cornell University Press Ltd., Ely House, 37 Dover Street, London W1X 4HQ.

International Standard Book Number 0-8014-1497-0
Library of Congress Catalog Card Number 82-7268

Printed in the United States of America

Librarians: Library of Congress cataloging information appears on the last page of the book.

The paper in this book is acid-free, and meets the guidelines for permanence and durability of the Committee on Production Guidelines for Book Longevity of the Council on Library Resources.

FOREWORD

It is a pleasure to write a foreword for this book, both because it honors Thomas Gold and because of the many fine articles contained in it. These articles give a picture of Gold's contribution to, and the current status of, three important areas of science. There are, however, many other areas on which he has had an impact—bringing innovation to all and controversy to many—and I want to offer some personal reminiscences (and irreverent observations) about a few of them.

I first met Tommy Gold in Cambridge, England, in the company of Hermann Bondi and Fred Hoyle around 1947. I knew they were working on something or other they called the steady state and sensed there was something sweeping about it, but I was then a young graduate student in quantum electrodynamics (high on mathematical rigor and experimental precision but low on speculation) and simply could not appreciate cosmology. At the same time, however, Tommy was also working on a resonance hypothesis for human hearing and *that* research made an immediate and indelible impression: his methodology was awe-inspiring (and almost frightening to a quantum electrodynamicist) in its combination of three seemingly incompatible ingredients—a willingness to question any basic principle (e.g.,

"Are servomechanisms generally important in physiology?"), an engineer's ability to analyze complexity concretely, and an interest in detailed evidence even if it were not quantitative ("were Beethoven's musicality and his particular kind of deafness related to the small change from a sharply resonant amplifier to an oscillator?"). I don't know whether modern physiologists quote Gold's early papers, but they now take some interest in feedback mechanism; they probably learned from Tommy more than they realize and less than they could have, had they listened to him more directly.

The magnetohydrodynamics of the solar system and of cosmic rays in the interstellar plasma ought to be related, but each field is so complex that, in the past, most workers specialized in one field or the other. Tommy Gold had the courage to work in both fields simultaneously—moving with ease from the corona of the sun to the corona of the galaxy—and some of that courage has rubbed off onto younger workers.

Sometimes even outdated papers have a positive effect later on, and this is especially the case with Gold's. The Gold-Hoyle model of a "hot universe" (Gold and Hoyle, 1959) is no longer compatible with modern X-ray observations for the universe as a whole, but their considerations of thermal instabilities are very relevant to modern X-ray data on the interaction of a galaxy with the intergalactic gas in a cluster. I sense that younger X-ray astronomers are insufficiently acquainted with this old paper. Switching fields as often and as radically as Tommy does is great fun, but one's earlier papers tend to be forgotten when one is no longer publishing in that field. Even though one is sometimes credited in the literature, the history of one's contribution is nevertheless foreshortened. There has, for example, been considerable interest in interstellar molecular hydrogen since the Copernicus satellite was launched. Though I am happy to say that modern authors on this subject often give credit to a detailed paper written in 1963 by Tommy Gold, Bob Gould, and myself, I must add that few people remember that Tommy had already written a short paper—without collaborators—on the same subject in 1961.

Even in the general areas covered by the articles in this book, there are some papers by Tommy Gold which, though they have not been quoted much lately, are becoming more relevant with the passage of time. For example, as observational cosmology turns more to questions of large-scale inhomogeneity and the borderline of open and closed universes, Gold's work on the arrow of time and on multiple universes (see entries 58, 77, and 139 in the Appendix to this volume) is of increasing interest. His papers on the expansion of relativistic gases (79) and on the evolution of dense star clusters (91) are two other examples.

A foreword about Tommy Gold would not be complete without at least a mention of controversies, and the "moon dust" story (see entry 41) is probably the best example. I am no expert in this field (nor do I have an ax to grind), but it seems to me that Tommy worked on microimpacts, plus cratering, plus gravity, plus electrostatic fields, and so on, at a time when other professionals were content to leave the lunar surface in its pristine state. I've always wondered why the lunar-science profession has given Gold a relatively "bad press," but I think it must be because he cannot resist making very detailed predictions, and the profession tends to judge on a linear instead of a logarithmic scale. A hypothetical professional might say, "Gold predicted the precise depth to which the astronaut's foot should sink in, but in fact the astronaut sank in to only 30 percent of that depth, which makes Gold 70 percent wrong (besides, he's wrong in calling it dust—we call it regolith, which sounds better!)." It seems to me, however, that a factor of 0.3 represents a rather small error on a logarithmic range of length-scales from one atomic monolayer on a pristine rock surface to the size of an astronaut's foot.

It is fascinating to watch another controversy unfold, namely Gold's ideas on methane in the deep earth's interior and on the primary causes (and the predictability) of earthquakes. I'm even less of an expert on this topic than on the moon, but I have a layman's intuition that history may repeat itself. The professionals in this field seem inclined to disregard gases in the deep interior and to say that earthquakes essentially cannot be predicted. If, thirteen years from now, Tommy should turn out to be qualitatively correct but the amount of nonbiogenic methane or the earthquake prediction should prove to be less than his most provocative predictions by a factor of 0.3, I leave it to the reader to fill in the remark of the 1995 professional! In case the reader is by now inclined to feel sorry for Tommy

Gold, it is important to remember that in 1995 Tommy is still sure to be making innovative hypotheses that make him the gadfly of yet another profession. I am certain he will continue to have fun (and so will the thoughtful 1995 reader), but in the meantime enjoy the articles in this book!

EDWIN E. SALPETER

Ithaca, New York

CONTENTS

CONTENTS

ILLUSTRATIONS

Editors' Preface

On 9–11 October 1980, many of Thomas Gold's friends and colleagues gathered at Cornell University in Ithaca, New York, to participate in a symposium celebrating his sixtieth birthday. The present volume is a collection of papers presented at that symposium. They were written by some of Gold's scientific associates and collaborators of long standing and by younger scientists who have been inspired by his ideas, teaching, and advice. The papers are grouped in three broad categories and to some extent reflect the breadth of Tommy's contributions to science, and to astronomy in particular. The first three chapters of Part I have a distinct character. The authors recall how they initiated and developed the steady state cosmology and discuss the consequences of the theory. These chapters also reflect on the human element that influenced and flavored science in the making.

Tommy has made important contributions in a great variety of subjects—ranging from the biophysics of human hearing to the nature of extragalactic radio sources—an achievement made possible by his unusual familiarity with a wide range of basic disciplines. (A bibliography of his scientific publications is provided as an Appendix to this volume.) As a young man Gold collaborated in the development of the steady state theory of the creation of the universe. Since then the focus of his scientific career has changed significantly a number of times but he has never ceased to approach problems from the broadest perspective.

When he was developing theories to explain phenomena such as the nature of pulsars, the mechanism of dust transport on the lunar surface, and the outgassing of the earth, his governing principle, we believe, was always to view each manifestation of the physical world as an integral part of the whole. For Gold, the explanation of a particular phenomenon is never an isolated task, a goal in itself, but rather an aspect of the total picture, consistent with a deep understanding of the basic laws of physics.

It is a pleasure to thank the authors of the essays included in this volume. In addition we express our appreciation to William A. Fowler, Raymond A. Lyttleton, John A. Wheeler, and Ronald Bracewell for their contributions to the symposium.

YERVANT TERZIAN
ELIZABETH M. BILSON

Ithaca, New York

PART I

COSMOLOGY

Plate 1. M81, a spiral galaxy. Courtesy, Palomar Observatory, California Institute of Technology.

1

Steady State Cosmology Revisited

Fred Hoyle

PART 1

The Steady State Theory, According to F. H.
(Hoyle, 1948)

Although the age of the whole universe was thought in 1948 to be only about 2 billion years, whereas the age of the earth was measured to be more than 3 billion years, this apparent contradiction did not worry me unduly, because there was obviously room for adjustment in Hubble's determination of the age of the universe. It was rather that I felt uncomfortable with an age of the universe generally comparable to the age of the solar system. I came to wonder if there could be any form of cosmology, based in a broad sense on Einstein's general theory of relativity, with the time-axis open into the past, not closed at the definite moment of the "origin" of the universe, as it is in the Friedmann cosmologies (Friedmann, 1922 and 1924).

The sudden creation in Friedmann cosmologies of all the matter in the universe also worried me. Indeed, it seemed absurd to have all the matter created as if by magic, as is still done today, amazingly in most quarters without a blush of embarrassment. I therefore began to see if the creation of matter could be put into a rational mathematical scheme.

The standard method in physics for going about such a problem begins with the construction of a suitable new field that contributes in a clearly defined way to the so-called action. There is then a standard mathematical process, known as the "principle of least action," which leads to gravitational equations similar to those of Einstein, but with an extra term in the "energy-momentum tensor," an extra term depending on the new field. I chose a scalar for my field, which became known subsequently as the C-field, and I constructed the contribution to the action from the derivatives of C, with the field taken to satisfy a coordinate-invariant wave-equation. This procedure fixed the situation uniquely within the framework of classical physics.

The consequences of these simple steps were startling. The solutions of the gravitational equations were quite unlike those obtained by Friedmann. The time-axis was indeed open into the past, as I had hoped it might be, and every solution of the equations tended asymptotically to a standard steady state form, a form with the metric

$$ds^2 = dt^2 - \exp(2Ht) [dr^2 + r^2 (d\theta^2 + \sin^2\theta \, d\phi^2)], \qquad (1)$$

and with a constant mass density ρ_0 related to the Hubble constant H by

$$\rho_0 = \frac{3H^2}{4\pi G}, \qquad (2)$$

where G is the constant of gravitation.

Unlike the Friedmann cosmologies, in which the Hubble "constant" is a misnomer since it changes from one moment of time to another, H was now a genuine constant, the same at all times. It was determined explicitly by the coupling constant of the C-field as the field appeared in the mathematical formula for the "action."

Equations (1) and (2) gave an expanding universe and yet with a constant nonzero mass density ρ_0, from which it was evident that matter had to be created continuously. There were objections from critics that the conservation of energy was being violated, but this was not so, because it is an important property of the mathematical scheme I have just described that it automatically guarantees the conservation of energy and momentum. This criticism is discussed at greater length in Part 3 of this chapter.

It was a more relevant objection that the theory, being classical, said nothing about the kind of matter that was being created. It could be hydrogen atoms, or carbon atoms, or blocks of soap. To theoretical physicists this omission seemed a serious defect, for they wanted to know the precise quantum details specifying the creation process. When I visited Zurich in the early 1950s, Wolfgang Pauli said, "If matter could be created it would be very good, but you must tell me exactly how it happens." If I could have answered Pauli to his satisfaction, everything would have seemed splendid. Yet an answer satisfactory in 1950 would have been quite unsatisfactory to theoretical physicists in 1982. The strength of a classical theory lies precisely in its ability to over-ride such details. Without this strength we would still know very little about gravitation.

The Steady State Theory, According to
Bondi and Gold (1948)

Hermann Bondi and Tommy Gold began their work from what is known as the cosmological principle, which states that an observer at an arbitrary point of space cannot distinguish his particular position from any other by making large-scale observations of the universe. Nor do the large-scale features of the universe show any difference between one direction and another. These properties describe the homogeneity and isotropy of space. Yet in the Friedmann cosmologies such an observer can distinguish the moment of his existence, because the large-scale features of the universe change with respect to time. Thus the Friedmann cosmologies are *not* homogeneous with respect to time.

What Bondi and Gold did was to postulate that the universe is homogeneous with respect to time as well as with respect to space.

From this hypothesis they were able to deduce that the space-time metric must have the form (1), but they were not able to obtain the mass density (2). Thus, the homogeneity postulate would have permitted the universe to be empty of all matter.

The Steady State Concept

The two points of view I have described have the affinity of both giving (1), arriving at this metric from the front and the behind as it were. They differ, however, in a crucial respect—in the meaning to be attached to the word "steady." All I could say from my point of view was that the universe had to be approximately steady with respect to the characteristic time scale H^{-1}. It was, on the other hand, in the very nature of the Bondi-Gold hypothesis that the universe had to be steady on a time scale much shorter than this, say on a time scale of $\frac{1}{100} H^{-1}$.

The Bondi-Gold point of view had drastic observational implications, since it permitted no property of the universe to change on a scale greater than about $\frac{1}{100} H^{-1}$, with respect to either time or space.* This was a far more disprovable position than my own, because much less distant observations than H^{-1} could be sufficient for establishing a disproof. It was therefore at the Bondi-Gold form of the theory that observational astronomers elected to shoot, and it was in these terms that all the arguments of the 1950s and early 1960s were conducted.

Astronomers were pretty well uniformly hostile to the theory. Their hostility was, at any rate in part, due to a paper written by Bondi in the early 1950s, in which he gave a list of past mistakes by observers. His conclusion, drawn from astronomical history, was that, in any face-off between observation and theory, it was theory that was more likely to come up with the puck.

The paper was submitted for publication to the Royal Astronomical Society. The Council of the Society was divided on whether the paper should be published with a majority favoring rejection. I happened at

---

*Here, as in (1), the units of space and time measurements are taken to be such that the speed of light is unity, $c = 1$.

that time to be a Council member, and I was therefore able to point out that Bondi's listing of observational errors were all taken from well-attested literature. "Was a paper to be rejected because its statements were correct?" I asked. This argument was having a little success around the table, when the president of the Society, W. M. Smart, squirmed in his Chair, and exclaimed in an anguished voice: "Then will somebody propose that this paper be rejected irrespective of its contents?"

Although the situation thus had its lighter moments, it was really quite badly one-sided. Besides being in a tiny minority of three, we then had no instruments for checking the statements of our opponents. Journals accepted papers from observers, giving them only the most cursory refereeing, whereas our own papers always had a stiff passage, to a point where one became quite worn out with explaining points of mathematics, physics, fact, and logic to the obtuse minds who constitute the mysterious anonymous class of referees, doing their work, like owls, in the darkness of the night.

Although I was not really as deeply involved as Bondi and Gold, since it was not my form of the steady state theory that was under direct attack, the observational claims were often so weak that I could not forebear speaking out against them. Weak they certainly were, as one can see from taking a look at them in a modern light. For brevity, I will confine myself to the two claimed disproofs that achieved the widest publicity in the 1950s and early 1960s: the red-shift magnitude relation for galaxies in clusters and the counts of radio sources.

The Red-shift Magnitude Relation

Figure 1.1 shows points for galaxies, published in 1978 by J. Kristian, A. R. Sandage, and J. A. Westphal (Kristian et al., 1978). The curve in this figure gives the red-shift magnitude relation for objects of the same intrinsic emission situated at different distances, calculated according to the metric (1). What is so very wrong with this, one may rightfully ask? In view of the scatter of the observational points, nothing much at all. Over the years, corrections applied to the actual observations have brought the corrected magnitudes into a reasonable correspondence with the steady state theory.

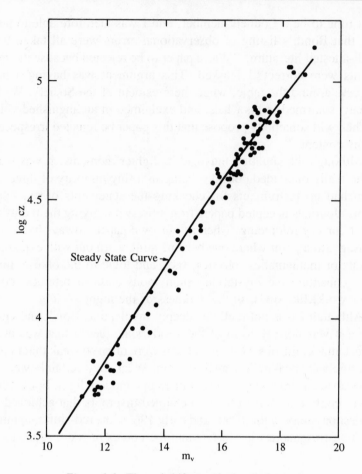

Figure 1.1. The red-shift magnitude relation.

Here, I must reveal that I have omitted three points from Figure 1.1, at the faint end of the magnitude scale. Since the observations stop at the faint end because reliability is being lost there, this omission cannot be seriously relevant to a test of the steady state theory. If one supposes the observers to have kept within a safe limit, and one gives equal weight to all their points, missing out three from so many can have no importance. It would be only if one wished to give excessive weight to what are likely to be the least reliable of the observations that the three omitted points would have relevance.

The Counting of Radio Sources

The observational procedure by which Martin Ryle claimed to disprove the steady state theory was as follows: Measure the flux received in an assigned bandwidth for all radio sources in a specified area of the sky that are within the confusion limit (or the sensitivity limit) of the radiotelescope. It is then a simple matter to count the number $N(S)$ with flux values larger than S. According to the Bondi-Gold form of the steady state theory,* the graph obtained by plotting $\log N$ against $\log S$ can never have a slope of magnitude greater than 1.5.

The observations were said to show a slope appreciably greater than 1.5, but the trouble with the observations was that just about every source of error one could think of would have the effect of falsely steepening the slope. This showed clearly in the results published from year to year. In the first Cambridge survey at 81.5 Mc/s the overall slope of the $\log N$-$\log S$ curve was said to be 3 (Ryle, 1955). When the frequency was raised to 159 Mc/s, reducing somewhat the effects of confusion, the slope became about 2.5 (Shakeshaft et al., 1955). Thereafter it became 1.8 (Ryle and Clark, 1961). Nowadays, using very high radio frequencies to avoid all confusion, the slope is so little different from 1.5 that the disproof of the steady state theory claimed in the 1950s and early 1960s fails completely (Wall and Cooke, 1975).

A different argument is used nowadays. It is said that the slope of 1.5 given by the Bondi-Gold theory is for sources not more distant than, say, $\frac{1}{10} H^{-1}$, whereas most of the observed sources are considerably more distant than this, in which case the slope according to the Bondi-Gold theory should be 1.2 or thereabouts. Hence, a contradiction occurs when the ground of the argument is changed.

The distances of most radio sources are not known, because for most of them no optically bright enough object has been found for a

---

*But not necessarily in the broader form of the theory, as Jayant Narlikar and I showed in 1961 (Hoyle and Narlikar, 1961).

red-shift determination to be made. If one confines attention to the minority for which distances are available and for which distances are indeed greater than $\frac{1}{10} H^{-1}$, the situation is just the way it should be, the $\log N$-$\log S$ slope is about 1.2 (Burbidge and Narlikar, 1976). All one therefore needs to do to defend the Bondi-Gold theory against this further assault is to deny that the unidentified sources are more distant than $\frac{1}{10} H^{-1}$, in which case they would need to be optically very subluminous.

I shall leave this defense, if they wish to make it, to Bondi and Gold themselves. Rather do I think it more interesting to explore the consequences of the unidentified sources being at distances appreciably greater than $\frac{1}{10} H^{-1}$. This would lead to severe difficulties, not just for the Bondi-Gold theory, but also for the Friedmann cosmologies. On the other hand, the broad form of the steady state theory, thinking of H^{-1} as the general cycle time for repeating operations in the universe, provides an interesting basis for resolving the problem.

In the Friedmann cosmologies, distances beyond $\frac{1}{10} H^{-1}$ also reduce the slope of the $\log N$-$\log S$ curve, due to the curvature of space-time demanded by Einstein's theory of gravitation. If the bulk of the sources are at such distances, how according to orthodoxy is the observed near -1.5 slope to be explained? By postulating that the sources were either intrinsically brighter, or more numerous—or both—at earlier times than they are today. We would need to suppose that such evolutionary effects compensate almost exactly for the curvature of space-time, so as to give the same results as if we were making observations for a nonevolutionary universe in Euclidean space.

Cyril Hazard put the situation very well when he said that, if historically the development of radioastronomy had come before the Einstein theory, all those astronomers who now accept this compensation as fortuitous would not have done so. Instead, everybody would have asserted that space was plainly Euclidean and that the Einstein theory was wrong.

The trouble in rejecting Hazard's comment with a faintly amused smile is that, according to the irrepressible optimism of popular opinion, a second similar fortuitous compensation is needed. If one counts optical quasi-stellar objects instead of radio sources, the distribution of number with respect to apparent magnitude is the same as if the observations were being made for Euclidean space and for a nonevolving homogeneous universe. Although this result has been known in a general way for ten years or more, it was confirmed in 1978 with great statistical weight by D. Wills and C. R. Lynds (Wills and Lynds, 1978).

This is at least one miraculous coincidence too many. I shall not discuss optical QSOs further, but for radio sources I am prepared to assert that there is no fortuitous compensation of the sort that is usually supposed. My view is that the unidentified sources emit their radio energies only over a short interval of cosmic time, so that we see them in a so-called time-hole,* all at effectively the *same distance* away from us. The log N-log S distribution refers to the intrinsic luminosity function of the sources, not to a space-time distribution. The need for any coincidence then disappears. Can we understand why the intrinsic luminosity function should give a 1.5 slope for the log N-log S distribution? The following explanation is for the well-known synchrotron process in which highly relativistic electrons (γmc^2 with $\gamma >> 1$) precess in a magnetic field.

Begin by considering a number of sources, all with the same initial electron distribution but with uniform magnetic fields of different intensities. An electron of energy γ in a field of intensity B can be taken to emit radiation with power $P = $ (known constant) $\gamma^2 B^2$, and with frequencies mostly close to $\nu = $ (another known constant) $\gamma^2 B$. Fixing the frequency (as is done in observation) permits B to be eliminated, to give an equation relating P and γ, $P = $ (known constant)/γ^2. The length of time for which each electron continues to emit radiation of frequency ν is proportional to γ/P, and hence to $P^{-1.5}$. It follows that the lower the value of P for the electrons in a particular source the longer they will continue to emit and the longer the source in question will continue to be a radio source. A future observer, living only a brief while, will therefore catch more sources with P low than with P high. Indeed, the number of sources the observer

---

*The distances are sufficiently the same for the observed fluxes to be dominated by the luminosity function and not by variations of distance.

counts varies with P as $P^{-1.5}$. For sources with the same number of electrons (at the same spatial distance from the observer) the dependence on P translates into a dependence on the observed flux S. The number of sources that are counted brighter than S is therefore proportional to $S^{-1.5}$, which is the required dependence.

Now add a second set of sources, occurring at the same cosmic epoch, and so observed at the same distance as the first set. Irrespective of whether the initial electron distribution is the same in the two sets, the observational result will be the same, except the counts fall to zero at different high values of S, thereby destroying the 1.5 slope at the highest fluxes when the results for the two sets are combined. A similar stochastic effect at the highest fluxes applies for a general distribution of initial electron distributions.

The lifetimes of the electrons in sources of sufficiently low emissivity (low magnetic field) may be so extended that the chance of the observer picking up such sources rises to unity. It is when S values become so small that an appreciable fraction of the counted sources have unit detection probabilities that the slope of the log N-log S falls off from 1.5.

The reader who examines the above argument in further detail will realize that the precise details of the log N-log S curve, especially the stochastic variations at high values of S, can be expected to depend on the observing frequency v, and this is exactly the behavior that is found in practice (Wall and Cooke, 1975).

What manner of cosmology could generate a profusion of radio sources at a particular epoch? The Friedmann cosmologies deliver matter in a veritable chaos that appears quite unable to explain such a precise property. A universe controlled by a strict cycle frequency is needed, analogous to a computer with an internal clock frequency. This idea will be developed later in Part 2.

The Microwave Background
in Anticipation

It often happens that a discovery of major importance might have been made long before it actually was. The discovery of the microwave background could conceivably have come as early as 1940, when Andrew McKellar observed a 3 K excitation of interstellar CH and CN molecules (McKellar, 1940).

George Gamow, from 1946 onward working in collaboration with R. A. Alpher and R. C. Herman, made the explicit choice of a Friedmann cosmology controlled in its early phases by the energy density of radiation, rather than by the energy density of matter (Gamow, 1946; Alpher and Herman, 1948; and Alpher et al., 1953). T_9 (in units of 10^9 K) was related to the time t (in seconds) elapsed since the origin of the universe by an equation

$$T_9 \simeq 10 \, t^{-\frac{1}{2}}.$$ (3)

The belief at first was that all the elements of the periodic table could be synthesized for values of T_9 between about 1 and 10. "The elements were produced," George said, "in less time than it takes to cook a dish of duck and roast potatoes."

In such a "hot big-bang" cosmology the radiation temperature continues to fall until the energy density of matter takes control, at which point equation (3) ceases to be valid. By the present day, the temperature would have fallen to a low value, and the 1948 estimate obtained by Gamow, Alpher, and Herman was 5 K (Alpher and Herman, 1948).

Interest in this point of view waned during the 1950s, as it eventually became clear that the elements, except possibly the very lightest ones, could not be synthesized in this way. In a lecture course given in January and February 1953 at the California Institute of Technology, I described what I believed to be a proof that galaxies could not condense in a hot big-bang cosmology. This old argument still stands today, and I will give it in a little more detail at a later stage.

For me, therefore, and I think for quite a number of others too, the hot big-bang cosmology was "out" during the 1950s. By 1963, however, I had returned to the ideas of Gamow, Alpher, and Herman, but with the interest shifted from synthesizing the heavier elements to synthesizing the lightest ones, helium especially.

If one looks back to the 1930s, 1940s, and 1950s, the last thing to worry astronomers was the synthesis of helium. It seemed good enough to know qualitatively that helium was being produced from hydrogen inside the sun and inside other stars. Only slowly did the quantitative aspects of helium synthesis take root. But by 1963 the quantitative problem of explaining the apparently universal abundance of helium relative to hydrogen was plain to be seen, and it was to this problem that Roger Tayler and I addressed ourselves in Cambridge, England. Essentially contemporaneously, P. J. E. Peebles was considering the same problem at Princeton University. Both these investigations postulated a hot big-bang cosmology, just as George Gamow had done. Results were published by Tayler and myself in 1964 (Hoyle and Tayler, 1964) and by Peebles shortly thereafter (Peebles, 1965), on the very threshold of the actual discovery in 1965 of the microwave background by A. A. Penzias and R. W. Wilson (Penzias and Wilson, 1965).

Later commentators have wondered why we did not use the observed cosmic helium abundance to predict what the present-day temperature of the microwave background had to be. Tayler and I avoided this inversion of the problem because the result depended far too sensitively on the precise helium abundance one happened to choose. If my memory serves me correctly, Peebles circulated a preprint in which he made such a calculation (using deuterium also) obtaining ~ 20 K for the background. I objected to this result because it contradicted McKellar's excitation temperature of 3 K—I was still too blind to see that McKellar had exactly the right number for the microwave background itself, not merely an excitation temperature.

Conformal Invariance and the
Conservation (or otherwise) of Baryons

There was a second reason, quite different from the helium problem, which caused me to stray away from the steady state theory. Already by 1963–1964, well before the actual discovery of the microwave background, I had become convinced that the equations of physics must not only be coordinate invariant but must also be invariant with respect to scale changes, invariant with respect to the group of conformal transformations. My trouble was that the wave-equation for the C-field, while being coordinate invariant, was not conformally invariant. To this I must add that for many years theoretical physicists had been insisting that the baryon number (essentially protons plus neutrons) was strictly conserved, a view that also helped to undermine my confidence in the steady state theory.

Yet over the years physicists have become more and more used to seeing their deeply cherished conservation laws going up in flames. A conservation law implies an invariance with respect to some group of mathematical transformations. To begin with, the group is thought to be complete, but then it is seen to be only a subgroup of a still wider group of transformations. Transformations going outside the previous subgroup are then said to "break the previous symmetry."

And so it has been with the conservation of baryons. It is a very major triumph of the steady state theory that the early challenge to it from theoretical physics has been swept away during the past five years. The current view is that the basic physical laws permit baryons to be created.

In just the same way I am happy today to accept that my previous insistence on conformal invariance could be too narrow a "symmetry." There could be wider mathematical transformations going outside the conformal group. Indeed, these former considerations now seem less relevant to me than the issues I will bring up in Part 2 of this chapter. I have mentioned my former difficulty, not because it is relevant to the situation of 1982, but because it explains why I was not in a sufficiently combative frame of mind to defend the steady state theory when the microwave background was discovered in 1965.

Before quitting the cosmological scene and leaving a younger generation to overgraze the infertile pasture of the big-bang cosmology, I did manage to get in a Parthian shot. If one took all the material in galaxies and spread it uniformly through space, the resulting average density would be $\sim 3 \cdot 10^{-31}$ g per cm^3, of which $\sim 8 \cdot 10^{-32}$ g per cm^3 would be helium. If the helium has come from hydrogen, the energy released per gram of transformed hydrogen being $\sim 6 \cdot 10^{18}$ erg, the average energy density released on a universal scale would be $\sim 5 \cdot 10^{-13}$ erg per cm^3, which is the same as the energy density of a microwave radiation field of temperature 2.8 K. Another fortuitous coincidence? For the past fifteen years, supporters of the hot big-bang cosmology have preferred to think so.

Thermalizing the Microwave Background

In a broad sense, then, a route toward defending the steady state theory against the criticism that it cannot explain the microwave background has always been open. Helium production would have to occur in stars, or in some other astrophysically locatable kind of object, and the energy released in the helium production would pour out into intergalactic space, initially in the optical, ultraviolet, and infrared regions of the spectrum. As I have just explained, this would give the correct energy density. It would then be necessary to thermalize the shorter wavelength radiation into the much longer microwaves.

The means of thermalization lay to hand, even as long ago as 1965, since by then we had essentially certain proof that about one-half of all interstellar carbon is condensed into graphite particles. This followed from the detection by rocket-borne equipment of exceedingly strong interstellar absorption centered around 2200 Å. Graphite has its strongest absorption over just this waveband, and there is no other kind of interstellar particle that can possibly have adequate abundance to explain the observed amount of the absorption. (It is relevant to this statement that the extinction near 2200 Å was found to be almost wholly due to absorption, not scattering.) The thermalizing particles would need to be in intergalactic space, however, as well as in galaxies. But there was no difficulty in seeing that explosions in galaxies could expel graphite grains into intergalactic space, as they are very likely doing in NGC 1275, for example.

Together with Chandra Wickramasinghe, Jayant Narlikar and Vincent Reddish, I tried this line of thought of course. For convenience in making calculations it seemed natural to choose our graphite particles to be spheres, and it was here that things went wrong, particularly as we left it to the Mie theory to churn out the results, instead of thinking physically about the problem. Mie theory showed that graphite spheres were hopelessly inefficient thermalizers, and, being in a preoccupied state of mind in 1965, I abandoned the argument at this point.

Tommy Gold was wiser. Instead of confusing himself with such calculations, he simply argued that "Nature" is almost always much more efficient at degrading nonthermodynamic situations toward the thermodynamic state than we expect. Provided "Nature" has been sly enough to make graphite particles of the right sizes and shapes,* Tommy can be seen to be correct.

For a graphite particle of volume V immersed in an oscillating electric field, the overall rate of energy absorption is $\sim\frac{1}{2}\sigma E^2 V$, where σ is the graphite conductivity and E the amplitude of oscillation of the field. This formula holds provided two conditions are satisfied. The relevant dimension or dimensions of the particle must be smaller than the so-called skin depth calculated from σ and from the oscillation frequency of the field, and the flow of current within the particle must not succeed in building electric charges at its surface which cancel the driving effect of the incident electric field.

The skin depth condition is no problem. It is indeed a help in balancing absorption at the shorter and longer waves, as will be seen in a moment. The condition on the building of free electric charges is crucial, however. It was just this effect that led to the negligible result of 1965. What happened in that calculation, concealed however within the Mie theory, was that surface charges built up at the longer wavelengths in only a small fraction of a cycle time, and the surface charges simply annulled the driving electric field, enormously reducing the electric currents flowing within the particles.

The skin depth of graphite at the shorter wavelengths is about 10^{-5} cm, which appears at first sight to restrict one to small particles, as in the 1965 calculation. But consider now a rod-shaped particle with radius small compared to its length, just as with the rods that make up a TV antenna. For the component of polarization of an incident field with electric vector along such a rod one now has the best of both worlds. It is, then, the small radius of the rod which matters for the skin depth, and it is the length of the rod which matters for the charge

*As was first emphasized by J. V. Narlikar, M. G. Edmunds, and N. C. Wickramasinghe in *Far Infrared Astronomy* (Pergamon, Oxford, 1976), p. 131. See also H. Alfvén and A. Mendis, 1977, *Nature 266*, 698.

separation. Indeed, adjusting the rod diameters at values around 10^{-5} cm provides a means for balancing the absorptive properties at the short and long waves—through the skin depth effect at the short waves. And by adjusting the lengths of the rods one can ensure that distant radiosources are not blotted out unduly at wavelengths longer, say, than 3 cm by an effectively thermalizing distribution of rod-shaped graphite particles in intergalactic space. The lengths of the rods required to give thermalization to a wavelength of 10 cm are of the order of 50 μm.

Returning to the simple formula for the average absorption by a particle, $\frac{1}{2}\sigma E^2 V$, and putting $\sigma \simeq 10^{15}$ sec^{-1}, a conductivity appropriate at the longer wavelengths, it is elementary to calculate that an effective thermalizer would require an integalactic density of rod-shaped graphite particles in the range from 10^{-34} g per cm^3 to 10^{-33} g per cm^3, a result that recently has also been obtained by N. C. Rana (1980). Since equation (2) gives $\rho_0 \simeq 10^{-29}$ g per cm^3 for $H = 50$ km per sec per mpc, the graphite density needs to be between 1 part in 10^4 and 1 part in 10^5 of the average universal density of the steady state theory. The needed fraction is satisfactorily low compared to the 1 part in $\sim 10^2$ of graphite in the interstellar medium within our own galaxy.

I have worked long enough among astronomers to be fairly sure that the typical reaction at this point will be to say, "I hope there is something wrong in all this." Bad luck. Then the typical reaction will be to say, "I hope that 'Nature' has not been sly enough to make this kind of graphite particle in appreciable abundance." Bad luck again. The shape of a carbon particle depends on the manner of its origin. If one puts a match to a strand of wool, it shrivels into a more or less spherical ball of soot. But if grains grow by sublimation from hot carbon gas, long rods—whiskers—are produced (Sears, 1955).

Whiskering occurs when gaseous atoms, adhering to the surface of a condensing particle, tend to run along the surface instead of being built immediately into the crystal structure of the particle. The incoming atoms run along the surface of a rod-shaped particle until they reach one or another of its ends, where, not being able to go any farther, they at last join the crystal structure of the particle.

Whereas a condensing spherical particle grows in radius only linearly with time, a whiskering particle grows in length exponentially with time. Eventually such particles break into pieces, and then each piece grows at an exponential rate. Further breakages occur, leading to a cascade of rod-shaped particles which grows exponentially in number, a catastrophic state of affairs. So there is nothing left to say, except, "I hope that somehow this doesn't happen in astronomy."

There is a seeming chance of better luck here. The graphite particles in the galaxy are certainly *not* long slender rods, and one might argue that graphite grains everywhere in the universe are the same as those of our galaxy. This form of uniformity postulate may possibly be useful in the absence of further knowledge, but in fact we do have further knowledge.

The situation for graphite grains in the galaxy is peculiar. Their lack of polarizing ability and their strong absorption and wavelength dependence combine to show that galactic graphite grains are small spheres with radii of only a few hundred Ångströms. The inorganic condensation of graphite from hot carbon gas produces either rod-like particles at low (astronomical) pressures or large, flake-like bits of soot at high (terrestrial) pressures. Inorganic condensation does not produce tiny spheres. For this result, one requires the degradation of an organic polymer. The degrading polymer shrivels into a small ball of carbon like a strand of wool in a flame.

I think it probable that essentially all the present carbon of interstellar space escaped from stars in the presence of an oxygen excess, and so formed into molecules of CO. Then H_2 was added to CO to give formaldehyde, H_2CO, and a polymer was constructed from a large number of units of H_2CO, $(H_2CO)_n$. If it happened subsequently that the polymer degraded, the result would be a small carbon sphere.

The condition for obtaining graphite rods is simply that carbon must be in excess of oxygen. While this is not the major present-day condition, the situation could have been otherwise in the early history of the galaxy, when a large fraction of all carbon was synthesized. The early carbon has now either quitted the galaxy (radiation pressure on a long slender rod being highly effective in causing expulsion into an intergalactic medium) or has been condensed into stars. Bad luck again, actually.

The Hot Big-Bang Does Not
Produce Galaxies

I turn now to my 1953 objection to the hot big-bang cosmology. There is no possibility in this theory of forming galaxies so long as matter and radiation remain coupled, because of the enormously high pressure of the radiation. Decoupling occurs at a temperature of about 3000 K and at a matter density of about 10^{-21} g per cm³. Condensation could then occur, if the matter were not in explosive expansion, provided the condensing gas remained essentially un-ionized and so avoided a destructive blast of radiation pressure. Ignoring the explosive expansion of the gas, let us suppose that condensation does take place. The Jeans length for a gas of temperature 3000 K and density 10^{-21} g per cm³ is about 10^{20} cm. Hence the resulting condensations would have a volume of order 10^{60} cm³ and a mass of about 10^{39} g. The condensations would therefore be of the scale of globular clusters, not galaxies, a result obtained repeatedly over the years by many investigators.

The more likely prospect is that nothing forms, for the whole material system is fragile compared to the radiation bath in which it is immersed, at any rate until the temperature of the latter falls below about 30 K. Nothing different from this position of 1953 has emerged during the past fifteen years, despite sustained effort by the whole of the world's astronomical community. Instead of the steady state theory being wrong because it cannot produce the microwave background, it is the hot big-bang that is wrong because it cannot make galaxies in a decisive way.

Helium Again

More needs to be said about helium. By 1960 it was already well recognized that the rate of helium synthesis inside stars is too small by an order of magnitude to explain a cosmic helium abundance equal to about one-third of the hydrogen, as observed in the sun, young stars, and the interstellar medium. Since it was unattractive to suppose these samples of material to be exceptional in their helium content, a serious problem for our galaxy had to be tackled—and also for other galaxies, assuming their material to be similar. It was just this problem which prompted the investigation in 1963–1964 by Tayler and myself, to which I referred above.

The possibility suggested itself quite insistently that most of the helium was really primordial. Everybody who tried making calculations for a hot big-bang cosmology, from Alpher, Follin, and Herman in 1953 (Alpher et al., 1953) to R. V. Wagoner, W. A. Fowler, and myself in 1967 (Wagoner et al., 1967), obtained a primordial helium abundance in good agreement with observation. The result was inevitable for matter that emerged from a really high-temperature radiation bath, provided only that the density of the emerging material was not too small.

White Holes in General and a White Hole of Galactic Mass in Particular

Although most of the galactic helium must surely have been formed in a high-temperature radiation bath, this does not prove the hot big-bang cosmology to be correct. The hot big-bang is sufficient for the origin of helium but not necessary. The hot big-bang is an example of a white hole, a white hole that is postulated to be so enormous that its products encompass everything we observe. But there is nothing to prevent us from thinking of a larger universe containing very many individual white holes, just as relativists have become accustomed to thinking of the universe as containing many separate black holes. Uncomfortably aware of this possibility, some relativists have invented excuses for why it could not be this way, but the excuses I have seen have been exercises in hand-waving, designed I would think to prevent one from proceeding along a road that might lead to unwelcome results. I propose here to follow that road.

Let me return for a moment to the position of 1948, to the broad form of the steady state theory based on a C-field. Earlier in this article I said that the C-field did not specify the precise form of the matter that it created. The created material could be hydrogen atoms, neutrons, or cakes of soap. Instead of cakes of soap, I am now going to suppose that creation takes place in the form of white holes. Such I will suppose to be the origin of galaxies, remembering that more than a decade ago Jayant Narlikar and I tried such an idea within the framework of the big-bang cosmology (Hoyle and Narlikar, 1966). Here I will consider the idea within the framework of the steady state theory.

Having decided that the baryon mass M of the white hole be of galactic order, say $M = 3 \cdot 10^{11}\ M_\odot$, where $M_\odot$ is the solar mass, the next question is the type of the white hole. Does it have insufficient energy to blow apart to a small matter density, in which case it will fall back into a black hole? Or does the white hole have an excess of energy over what is needed for complete expansion, in which case it will appear as a giant explosion emerging into the outside world? Or is the energy intermediate, sufficient for the white hole to expand apart to a diffuse state, but not sufficient to present the aspects of a really violent explosion? These three types of white holes are known technically as elliptic, hyperbolic, and parabolic. They correspond to the types $k = 1$, $k = -1$, $k = 0$ respectively in the Friedmann cosmologies.

Observation suggests that all three cases occur. The explosions of QSOs and radio galaxies suggest an association with the hyperbolic type ($k = -1$). The bulk of the material of normal galaxies would seem to correspond to the nearly parabolic case, while at the very centers of galaxies there are compact massive objects naturally to be associated with the elliptic type ($k = 1$). These massive objects are probably composed of material that lacked the energy to join the much more extended outer regions of the galaxies, and which fell back into black holes.

For the material forming the bulk of a normal galaxy we require the parabolic case, which limits the energy of the radiation bath in which the material emerges from its white hole. By "emerge" I mean the crossing of a sphere of surface area $4\pi R^2$, the hole being taken as spherically symmetric, where R is related to the total energy at emergence by

$$R = \frac{2G}{c^4}\ \text{(Total Energy at Emergence)}, \qquad (4)$$

c being the speed of light.

At emergence, the radiation bath must not have appreciably more energy than Mc^2, otherwise we would have a hole of the hyperbolic type. Writing ρ_R, T_R at emergence, and putting the two energy contributions equal in order to define the parabolic case, we have

$$\rho_R c^2 = aT_R{}^4 = \frac{1}{2} \text{ (Total Energy at Emergence)}, \tag{5}$$

$$M = \frac{4\pi}{3} \rho_R R^3, \tag{6}$$

where a is the Stefan constant. For $M = 3 \times 10^{11} \odot$, equations (4), (5), and (6) solve easily to give

$$R = \frac{4GM}{c^2} = 1.78 \times 10^{17} \text{ cm}, \tag{7}$$

$$\rho_R = \frac{3M}{4\pi R^3} = 2.54 \times 10^{-8} \text{ g per cm}^3, \tag{8}$$

$$T_R = \left(\frac{\rho_R c^2}{a}\right)^{1/4} = 7.41 \times 10^6 \, K. \tag{9}$$

A scale factor S can be used to define various stages in the expansion of the white hole, with $S = 1$ at emergence through the sphere of radius R. Thus the proper area of the outer boundary of the object at stage S is $4\pi R^2 S^2$. The mathematical equations of general relativity require S to change as $t^{1/2}$ inside R, but as $t^{2/3}$ when S increases above unity. Here t is proper time measured by an observer moving with the object. This behavior of $S(t)$ is exactly similar to the red-shift scale factor in the hot big-bang cosmology. The following proportionalities are also the same as in the hot big-bang cosmology

$$T \propto S^{-1}, \tag{10}$$

$$\rho \propto S^{-3}, \tag{11}$$

with (10) holding for the temperature of the radiation until it eventually uncouples from the matter, and with (11) determining the variation of the baryon density.

When the matter, composed overwhelmingly of hydrogen and helium, no longer maintains a sufficient degree of ionization to hold the radiation, uncoupling occurs. This uncoupling happens when T falls to ~ 4000 K, a rather higher value than in the hot big-bang (because the object is of finite size and because the density values are higher at corresponding values of T). Explicitly, the numerical values at the uncoupling of matter and radiation are

$$T \simeq 4000 \text{ K,} \tag{12}$$

$$S \simeq 2000, \tag{13}$$

$$\rho \simeq 4 \cdot 10^{-18} \text{ g per cm}^3. \tag{14}$$

White Holes in the Steady State Theory

Unlike the situation that obtains in the hot big-bang theory, a localized white hole loses its radiation as soon as decoupling occurs: the radiation simply streams out of the material once its opacity disappears through the drying up of ionization of the hydrogen. Also unlike the case in the hot big-bang theory, a localized white hole loses neutrinos, a process that occurs long before the radiation escapes.

Neutrinos are generated inside a white hole at very high temperatures, above 10^{10} K. They decouple from the matter already at

$$T \simeq 10^{10} \text{ K,} \ S \simeq 1/1300, \tag{15}$$

far inside the radius R. Because the neutrinos have at most only a small rest mass they move essentially at the speed of light, which permits them after decoupling to expand faster than the rest of the white hole. Although the neutrinos start their escape at $S \simeq 1/1300$, a reasonably straightforward calculation shows that on the average they do not manage to reach the outer boundary of the object until $S \simeq 1/4$.

The energy density of the neutrinos at $S = 1/4$ is closely the same as the energy density of the radiation, which at $S = 1/4$ is 4 times the energy density of matter. These statements follow from (5), (10), and (11).

The emerging neutrinos have about 4 times the rest mass energy of matter for an observer comoving with the matter, which does not tell us what energy the neutrinos will have when they eventually arrive in the distant universe. An additional factor of transformation between the comoving observer and a distant observer must still be included.

Rather surprisingly perhaps, the additional factor is one of blue shift, as was shown in 1964 by John Faulkner, Jayant Narlikar, and myself (Faulkner et al., 1964). Our conclusions have recently been reconfirmed and extended by Narlikar and Apparao (1975).

The blue-shift factor depends on the value of S at which the neutrinos reach the surface of the object. For $S = 1/4$ the factor is quite modest, about 2.5. Emergence at much smaller S would give a far larger factor, as can be seen from the results given in the papers just cited.

The upshot is that the neutrinos emerge into the distant universe with about 10 times the mass energy of the hydrogen and helium delivered by the white hole. Hence it follows that, if white holes of galactic mass are responsible for maintaining the steady state theory value of ρ_o given by (2), neutrinos must make the main contribution to ρ_o. With $H = 50$ km per sec per mpc, we have

$$\text{Neutrino mass density} \simeq \rho_o = \frac{3H^2}{4\pi G} \simeq 10^{-29} \text{ g per cm}^3, \tag{16}$$

$$\text{Hydrogen plus helium rest mass density} \simeq \frac{1}{10} \text{ (Neutrino mass density)}$$

$$\simeq 10^{-30} \text{ g per cm}^3. \tag{17}$$

The kinetic energy of an individual neutrino averages about 10 kilovolts, although the comparatively few neutrinos that reach the boundary of the object for values of S considerably smaller than one-quarter have kinetic energies that may be as high as 1 million electron volts, or more in a few cases.

The Microwave Background Again

From (5), (10), (11), and (12), it follows that the energy density of the radiation escaping from the white holes must be about 1/2000 of the rest mass energy of the hydrogen plus helium. The latter is obtained simply by multiplying (17) by c^2. Hence we get

$$\text{Average universal energy density of radiation} \simeq \frac{10^{-9}}{2000} = 5 \cdot 10^{-13} \text{ erg per cm}^3, \tag{18}$$

in agreement with the energy density of the microwave background.

At escape from the white holes the radiation becomes a dilute blackbody distribution of temperature ~ 4000 K. To correspond with observation, thermalization toward a nondilute blackbody distribution at temperature ~ 2.8 K is required, taking place in accordance with the properties of needle-shaped graphite particles.

In the broad picture of a universe that is steady with respect to the generation time H^{-1}, we can think of a thermal background of temperature, say, T_b, produced in some phase range during each generation. For the current generation the background would still have temperature ~ T_b, but for the preceding generation the red-shift effect would by now have reduced the temperature to ~$e^{-1} T_b$, and to $e^{-2} T_b$, $e^{-3} T_b$, . . . for still earlier generations. The observed microwave background would therefore be a superposition of blackbody distributions with temperatures T_b $(1, e^{-1}, e^{-2}, . . .)$. Only the current generation would contribute at the highest frequencies, whereas several generations would add together at low frequencies on the Rayleigh-Jeans "tail."

Recent data obtained by D. P. Woody and P. L. Richards (1979) show this expected effect, with the highest measured frequencies indicating a lower temperature, ~ T_b, than do the lower frequencies, ~ T_b $(1 + e^{-1} + . . .)$. The present discussion is of course much simplified, so that precise numbers should not be pressed too far. Figure 1.2 shows the observational results.

The Helium Abundance (for the last time) and the Deuterium Problem

From (7), (8), (10), and (11), which are applicable so long as the scale factor S is less than ~ 2000, one can write

$$\rho = 6.24 \times 10^{-2} T_9^3, \tag{19}$$

where T_9 is the radiation temperature in a white hole of galactic mass in units of 10^9 K. The factor 6.24×10^{-2} is the parameter h_0 of Wagoner, Fowler, and Hoyle (1967), and also of Wagoner (1973). Tables given by these authors show that the mass fraction of helium in emerging material with $h_0 = 6.24 \times 10^{-2}$ must be about 30

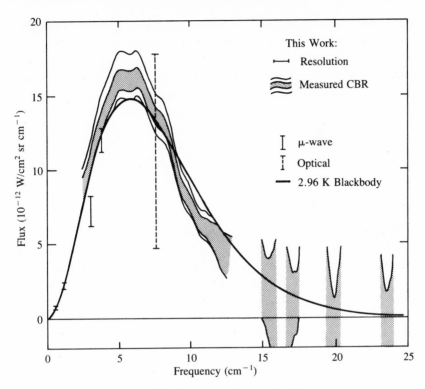

Figure 1.2. The observed microwave background.

percent. This accords very well with the usually quoted helium values for young stars and for the interstellar medium.

Supporters of the hot big-bang cosmology will have noticed that the present model does not provide a significant direct yield of deuterium—let them derive comfort from this fact if they please. Our knowledge of the existence of D is fragmentary. While I would not seek to argue that the occurrence of D is limited to the few places where it has been detected, I would argue that these are just the places where the concentrations of D happen to be unusually high. Thermal effects must tend to produce marked concentrations, because of the large two-to-one mass ratio of D to H, and observation must tend to pick out those special regions where the factors of concentration have become large.

At $h_0 = 6.24 \times 10^{-2}$, the results of Wagoner, Fowler, and Hoyle gave a ^{7}Li mass fraction of $\sim 10^{-6}$. Two neutrons and a proton in ^{7}Li are only lightly bound, and combinations of these three baryons are easily shaken loose by nonthermal collisions. We must even contemplate the spallation of ^{4}He. Although the spallation of ^{4}He needs more violent collisions than ^{7}Li, the abundance of D is so low in comparison with ^{4}He that one would need great confidence in one's knowledge to be sure that sufficient collisions on ^{4}He had not occurred. This possibility is enhanced by the following further details.

A Hierarchical Picture

For simplicity I have considered just one white hole in isolation from all others. The clustering of galaxies suggests a more complex situation, however, as if there had been a hierarchy of white holes, with smaller white holes emerging from larger ones. Such a situation would be very productive of shock waves, since matter from one white hole would tend to impinge on matter from others, a circumstance of relevance to the spallation issue discussed in the preceding paragraph.

This more realistic picture, while retaining the same broad features, would change the numbers in the detailed manner given later in the Appendix. White holes yielding clusters of galaxies would have masses $\sim 10^{14} M_\odot$, for which the parameter h_0 would be lower than the value given in (19). The effect would be to reduce the amount of ^{4}He by a few percent, and also to provide more radiant energy for the microwave background than (18) does. Since the latter has probably experienced some reduction in its energy density through the expansion of the universe, such an augmentation of the radiant energy would not be a disadvantage.

If one looks at the universe empirically, without being nagged by conventional cosmological ideas, or by a half-knowledge of the physical laws, this is exactly the way it looks. More than twenty years ago this was the way things appeared to V. A. Ambartzumian (1958). Ambartzumian's ideas did not have very much impact on contemporary thinking, essentially because they did not accord with either the cosmological or physical ideas of the day. Rereading in 1982 his paper given at the XIth Solvay Conference (1958), Ambartzumian's contribution appears remarkably percipient.

Although I vowed at an earlier stage to say nothing more about QSOs, I cannot resist just one remark in relation to the present hierarchical system of white holes. So far, the picture has been considered parabolic, with the emerging material having just sufficient dynamical energy to form a bound cluster of separate objects. In a hyperbolic case the objects would continue to fly apart instead of forming a bound cluster, and in a violently hyperbolic case the velocities of separation of the objects would be large.

By now too many close pairs of QSOs have been discovered for their associations to be due to chance. The different red-shift values of members of such pairs raise a problem that might find its solution in terms of a violently hyperbolic system of white holes. It is possible that we do not observe by any means all of the members of such systems. Most of them may emit too little light for optical detection to be possible. If the direction of motion is considered to define a polar axis, and if the emission is nonisotropic, with marked concentration toward the rearward pole QSOs would be seen only in tranverse motion. Tranverse motions always produce red shifts.

Condensations within a
Galactic White Hole

When radiation decouples from matter at the temperature and density values given in (12) and (14), the radiation streams out of the matter, leaving it free to condense. The situation is not plagued by the continued presence of the radiation, as it is in the hot big-bang cosmology. Since we have been contemplating a parabolic situation for the emerging matter, regions where the density is sufficiently higher than the average will have gravitational fields that are strong enough to prevent more than a modest degree of further expansion of the material, which will then contract and fragment into masses determined by the Jeans length. What are these masses?

The Jeans length for $T \simeq 4000$ K, $\rho \simeq 4 \cdot 10^{-18}$ g per cm^3, as in (12) and (14), is about $3 \cdot 10^{18}$ cm, and the mass contained within a sphere of this radius is about $3 \cdot 10^4 M_\odot$, at the low end of the range of globular cluster masses. Since those globular clusters that have survived for a time H^{-1} as separate entities are likely to be condensations with above-average masses, this result again accords well with observation.

It is a further advantage of the present theory that the material emergent from a white hole of galactic mass has a sufficient content of carbon and nitrogen to operate the CN-cycle already in the first stars (cf. table of abundances in reference 21). There is no awkward hiatus in the structure and evolution of stars as there is in the hot big-bang cosmology, for which the amounts of primordial carbon, nitrogen, and oxygen are negligible.

PART 2

The Information Content of the Universe and Its Relevance for Cosmology

I believe that the most important datum for cosmology—more important than any of the astronomical details discussed above—is the very large information content of the universe.

The information shows itself quantitatively in the enzymes, for example. An enzyme is a protein with a very particular surface shape that attaches to itself the constituents in a biochemical reaction, bringing them into close propinquity. Thus jammed together, as it were, the constituents react with a probability far higher than would otherwise be possible—in the general melee of a water solution, for example. When the reaction is completed, the product(s) leaves the enzyme, which then repeats the process.

Surface shape is therefore all-important to the function of an enzyme. Surface shape is determined by the particular sequence of amino acids in the polypeptide structure. One can think of getting the surface shape right in two stages of approximation. There are ten to twenty distinct amino acids that determine the basic backbone of the enzyme, and these amino acids simply must be in the correct positions in the polypeptide structure. The rest of the amino acids, usually numbering one hundred or more, then control the finer details of the surface shape. At this second order of approximation the detailed positioning of the amino acids can be varied considerably.

Consider now the chance that in a random ordering of the twenty different amino acids that make up the proteins, the different kinds fall into the order appropriate to a particular enzyme. The chance of obtaining a suitable backbone can hardly be greater than 10^{-15}, and the chance of the rest being tolerable to the functioning of the enzyme is also small. I doubt I will be accused of setting the overall probability too high if I calculate with a chance of 10^{-20}.

By itself, this small number can be faced, because one must contemplate, not just a simple shot at obtaining the enzyme, but a very large number of trials, such as is supposed to have occurred in an organic soup early in the history of the earth. The trouble is that there are more than two thousand different independent enzymes necessary to life, and the chance of getting them all produced by a random trial is less than $10^{-40,000}$. This minute probability of obtaining all the enzymes, only once each, could not be faced even if the entire universe consisted of organic soup.

The same enzymes appear across the whole face of biology: they are the same in man as in a yeast cell or a bacterium. Nothing in the evolution of life on the earth has ever shifted the basic backbones of the enzymes. This remarkable fact gives a strong hint that life is a cosmic phenomenon, while the number $10^{-40,000}$ shows that within the context of the whole universe something very crucial was required for their origin.

There are other biomolecules for which similar calculations can be made. Indeed the histones and t-RNA molecules are more strictly invariant across the whole of biology than the enzymes. For every such set of biomolecules the chance of a random discovery is exceedingly small, so very small as to wipe entirely out of court the idea of producing life without a large initial supply of information.

It is idle to argue that the information lay in a sterile physical environment at the surface of planet earth in the early days of the solar system. A somewhat better idea is to suppose that the information lies in the physical laws themselves, essentially in the Schrödinger equation. Some years ago, I used to think this idea could be the only possible solution to the dilemma. I then became suspicious that it was merely a notion drawing whatever strength it had from our inability to work out the detailed consequences of the Schrödinger equation.

Suppose one contemplate a silicious soup instead of a carbonaceous one. Why not have the Schrödinger equation leading inevitably to the silicon chip and an organization of computers? Every stream and every lake is an actual silicious soup, but nothing much ever happens in them except the precipitation of simple crystals of silica. This analogy suggests that the evolutionary notion is absurd, particularly as the information content needed in the Schrödinger equation to produce a computer would be slight compared to the information needed for the origin of carbonaceous life.

When it became apparent to me that life is an extraterrestrial phenomenon, the consequent broadening of the environment came as a short-lived relief. But then it soon became clear that extending the environment in a purely physical sense simply did not touch the problem. There is no way to cope with $10^{-40,000}$ except through an enormous broadening of the information supply. The question was, where in the universe would there be adequate information? Not, I decided, in a big-bang cosmology. The only possibility for accumulating the needed information seemed to lie in opening the cosmic time-axis into the past, exactly my old starting position of 1948.

The problem for a big-bang cosmology is acute, especially if one moves to the higher values of H that are now coming into vogue. With $H = 100$ km per sec per mpc, a value currently popular, the age of the universe at the closure density of the Friedmann cosmologies, $3H^2/8\pi G$, is less than 7 billion years. Life existed already in a comparatively sophisticated form on the earth almost 4 billion years ago (Phlug and Jaeschke-Boyer, 1979).* This fact leaves a mere 3 billion years for the accumulation of the vast information content needed to produce the enzymes, and to do so, moreover, out of chaos—the chaos is so great that the big-bang cosmology cannot even produce a galaxy, much less a living cell.

---

*Professor Phlug has considerably extended the work described in this reference, to a point where the existence of fossilized yeast cells in the Greenland Isua rocks has been put beyond doubt.

There are other issues of a broader kind that were noted already in the 1950s, at the time of the early investigations of stellar nucleo-synthesis. Is the positioning of the level at 7.65 Mev in ^{12}C an accident? Is it an accident that the 7.12 Mev level of ^{16}O lies just below the sum of the rest masses of ^{12}C and ^4He? Without these circumstances together, the cosmic ratio of C to O could not be appropriate to life, which demands approximately equal abundances of these two crucial elements.

The sum of the masses of the proton and electron is fractionally a little less than the mass of the neutron. The reader may like to contemplate the kind of universe we would have if the situation were the other way around. The student of stellar structure may like to consider what would have happened if ^8Be had been slightly stable rather than slightly unstable. A fuller compilation of these "anthropic" issues was given in 1979 by B. J. Carr and M. J. Rees.

Modern developments in particle physics, by destroying the tidy 3-quark world of the 1960s, have made it a little less difficult to conceive of an explanation for these broad features, so overwhelmingly important for our existence. We can think of both experimental physics and astronomy as being manifestations of only a subset of the possible states of the complete universe, a subset in which the so-called coupling constants of physics happen to take on the particular values that explain the apparent accidents mentioned above. No terrestrial experiment or observation goes outside this subset of states, which can be regarded as physical information, to be considered together with the information necessary for life. My concept is of a universe evolving slowly with respect to its information content. Let me explain this process in general terms using a few everyday examples.

Almost every efficient human information-carrying device is associated with the modulation of some cyclic system. We broadcast information by radio using either the modulation of the amplitude or the frequency of an electromagnetic carrier wave. Access to information in a computer depends crucially on the timing cycle of the computer. Likely enough too, there is a basic cycle frequency in our own brains. The cycle need not be of an electrical nature. A similar idea applies to the playing of a phonograph record, with the mechanical rotation of the turntable supplying the cycle. Speech depends on the modulation of sound waves in the air.

These examples make it easy in returning to (1) to think of H^{-1} as the basic cycle time of the universe. Just as many cycles are needed to develop much of a message in the above human cases, so many cycles of the universe have been necessary to develop the information needed for life, and needed for the specification of the physical constants as we find them.

There is no sense in which I would think it helpful to return to the concept of a universe that is strictly steady, not even on the scale H^{-1}, and certainly not on the much shorter time scale of $\frac{1}{100} H^{-1}$. My 1948 form of the steady state theory was rather like a phonograph record stuck in a groove, playing endlessly the same phrase, while the Bondi-Gold form of the theory was like a record that plays only one note, the sort of test record one uses to check a stereo system.

One does much better to listen to a recording of Beethoven's Fifth Symphony. If one analyzes the music with each rotation of the turntable, one sees that nothing much usually happens in a single cycle (although occasionally there is a sudden outburst when something drastically different really does happen). Over the course of an entire movement, however, involving very many rotations, the information content of the music does change—the more so if we substitute a late quartet for the Fifth. Yet from beginning to end throughout the movement there is a coherence to the structure of the music, analogous to the working through of the consequences of the values of the physical coupling constants that specify a particular subspace of the universe. Changing the subspace would be like changing the record from Beethoven to Mozart.

Possibly a reinterpretation of the steady state theory along these lines will receive a more sympathetic reception than did the theory of 1948. In retrospect, I can see that not everybody welcomed the idea of listening to a record that had become stuck in a single groove. Nor does it take forever to test out a stereo system. In retrospect, it is a pity that it was not possible to break loose from the restraining grip of observational astronomy, which forced one in 1948 to avoid any suggestion of times and distances greatly exceeding H^{-1}. Nobody was comfortable with more than $\frac{1}{3} H^{-1}$, only a third of a single rotation of the turntable. Experience shows that you can't play much music in that short a time.

Yet in giving universal significance to the Hubble constant, the carrier frequency, the steady state theory of 1948 achieved a breakthrough that seems more and more relevant as the years go by. It was unfortunate that in 1965 those of us who had either worked with the theory or who viewed it with some sympathy allowed ourselves to be bamboozled into impassivity, thus permitting the astronomical world to plunge with avidity into what has come increasingly to look like a blind alley.

Finally, a return to (1) and (2) provides a complete absorber along the future light cone, and so permits the generation of all physical fields to have strict time-symmetry (Wheeler and Feynman, 1945; Hoyle and Narlikar, 1974). The time sense of the universe becomes unique, another telling and perhaps crucial point in favor of the steady state theory.

PART 3

A Historical Note

I returned in 1945 from wartime activities to a temporary post in Cambridge. The house my wife and I rented was so far out in the remote countryside that from Monday to Saturday morning in university terms I mostly found it convenient to live in St. John's College.

Hermann Bondi had rooms in Trinity, about halfway along the arm of Great Square nearest to Trinity Street. Often I would call on him, particularly when in the period 1945–1947 he became interested in the theory of relativity. Einstein had modified his 1915–1916 form of the theory by introducing into his equations a so-called cosmical constant, which had the effect of requiring particles to repel each other when separated by large distances. George Lemâitre had used Einstein's modified theory to invent a cosmology which he called the "primeval atom," a cosmology different from those obtained in 1922–1924 by Alexander Friedmann. Lemâitre's discovery, which had been strongly supported by Arthur Eddington, showed that Friedmann had not exhausted all the cosmological possibilities, a circumstance that opened our minds to new ideas.

Einstein's modified theory still required particles to attract each other gravitationally at distances on the scale of an individual galaxy, however. Bondi and I began our own thinking by wondering what would happen mathematically if particles repelled each other at all scales of distance. We soon ran into trouble and dropped the notion, but it served to familiarize us with the mathematics of cosmology. Bondi continued his work by writing a Council Note dealing with the subject for the Royal Astronomical Society—a note that he later expanded into a Cambridge Monograph.

An important thread for us was the Dirac large-number hypothesis. Very large numbers appear in physics, as for instance when one takes the ratio of the electrical and gravitational forces between an electron and a proton. It had been known for about two decades that these very large physical numbers were uncannily close to other very large numbers arising from the structure of the universe. The latter changed with time according to the cosmological theories of Lemâitre and Friedmann. Dirac, having rejected the possibility that the physical and astronomical coincidences were fortuitous, had therefore argued that if the cosmological numbers change with time then so must the physical numbers. In particular, the "gravitational constant" of Newton (also appearing in Einstein's theory) had to change with time, a circumstance of relevance to a host of astronomical problems, for example, to the orbits of the planets and to the luminosity of the sun in bygone aeons. Such a change also affects the tests of Einstein's theory, causing serious trouble when the correspondences of theory and observation are tightly pressed.

With this background, I will take up the story of the steady state theory on the night in 1946 when Hermann Bondi, Tommy Gold, and I went to see a ghost-story film, which had four separate parts linked ingeniously together in such a way that the film became circular, its end the same as the beginning. I have not been able to trace the name of the film but, drawing on a remote corner of my memory, I think it was called *The Dead of Night*. Tommy Gold was much taken with it and later that evening he remarked, "How if the universe is constructed like that?"

One tends to think of unchanging situations as being necessarily static. What the ghost-story film did sharply for all three of us was to remove this wrong notion. One can have unchanging situations that are dynamic, as for instance a smoothly flowing river. The universe had to be dynamic, since Hubble's red-shift law proved it to be so, but if the universe could be unchangingly dynamic, like a flowing river, the very large numbers of cosmology would also be unchanging. Hence by Dirac's argument the very large physical numbers would not change either, and the gravitational constant in particular would be genuinely constant and there would then be no problems for Einstein's theory to worry about.

From this position it did not take us long to see that there would need to be a continuous creation of matter. As galaxies expanded apart from each other, new ones would have to be born in the spaces that opened up in order to preserve an unchanging situation. This process would eventually exhaust the available matter in space unless new stuff was created. At this point we made the grievous mistake of thinking the creation of matter would violate conservation of energy, and because of this erroneous belief we came to a dead stop. Nothing further was to happen until the end of 1947.

In late December 1946 I was asked to give a paper at a meeting of the Physical Society of London, a meeting actually held in Birmingham. I made the journey on a very cold day, driving a small ancient car as far as Northampton, and then taking an almost equally chilly train the rest of the way into Birmingham. My subject was: "On the Formation of Heavy Elements in Stars." During the discussion period, Rudolf Peierls made the following interesting remark:

> Most of the discussions of abundance seem to start from the hypothesis that the primary element is hydrogen and in some stage the universe consisted of hydrogen only. This is the simplest, but by no means the only, possible hypothesis, and if we explain the origin of heavy elements by their formation from hydrogen the next question is evidently, where does the hydrogen come from?

So here I was, back at the creation of matter, this time from a physical, rather than a cosmological point of view. I had much else to do throughout the spring of 1947, and it was not until the long summer break at Cambridge that I was able to turn to the problem of the creation of matter, and to the issues I described at the beginning of Part 1.

It was not until November 1947, however, that at last I saw the conservation of energy objection had been wrong. We had been thinking in the previous year of the situation as if it were taking place inside a closed fixed box, but where for the universe was there a closed fixed box? One possibility was to define a box kinematically, with its boundary determined by a set of galaxies of fixed identity. If one did this, then energy was conserved in the cosmologies of Lemâitre and Friedmann. But how if one defined a box with fixed rulers? Then energy was not conserved in the cosmologies of Friedmann and Lemâitre. For energy to be conserved within a box of fixed "proper" volume one would need an unchanging universe with creation of matter. And why should a box defined with fixed rulers not be just as good as a box defined by galaxies of fixed identity? Why not indeed.

At this point it became clear that there was no further profit to be gained from such a qualitative style of argument. I had to turn to the mathematical definition of energy conservation, meaning that I had to keep to the logical structure of Einstein's general theory of relativity, which equates a well-defined geometrical quantity to a physical quantity. There was a popular belief as to what the physical quantity should be, but there was nothing at all in the theory which required the popular belief to be true. Consistent with a certain quite general mathematical rule (that it should be a symmetrical, second-rank tensor), I could modify the physical quantity to please myself. The clear question was: Could the physical quantity be modified so as to produce creation of matter, and if this could be done would the universe turn out to be dynamic but unchanging?

It was in January–February 1948 that I found positive answers to these questions. I wrote a paper entitled "A New Model for the Expanding Universe," which I presented at a seminar given in the Cavendish Laboratory on 1 March. Both Paul Dirac and Werner Heisenberg were there, and afterward I heard Heisenberg had returned to Germany with the comment that the seminar was one of the interesting things he'd heard during a six-month stay in Britain.

My mathematics was then just about good enough to answer the questions I had asked. It was to be improved in 1951 by M. H. L. Pryce, whose formulation of the problem led to the same broad conclusions as mine, which was mainly why he never published it—a decision characteristic of Maurice Pryce, who published only about one-tenth of what he should have done.

In early March 1948 I submitted my paper to the Physical Society of London, choosing the Physical Society because its meeting in December 1946 had triggered the critical phase of my thinking. On 10 May I had a reply from the Papers' Secretary in which the following was the critical paragraph:

> I apologize most sincerely for any inconvenience you may have been caused through the delay in receiving information on your paper on "A New Model for the Expanding Universe." Your paper has had very serious consideration by the Papers Committee of the Society who have now regretfully decided that the Proceedings is not the most suitable medium of publication, especially in view of the acute shortage of paper which is forcing us to reject papers we would otherwise be glad to publish. The Committee suggests that you submit the paper to the Royal Astronomical Society.

Although this reply was courteously worded, I felt that a little more weight might have been given to my efforts on the society's behalf in the previous December. Unfortunately, the Royal Astronomical Society in 1946–1947 had been taking eighteen months to publish papers. Not wishing to wait so long, I sent the new steady-state theory to the *Physical Review* in the United States. By the end of June I had a reply, offering publication if the paper were cut to half its length. The paper was not particularly long, however; indeed, I had already made it as short as I thought possible consistent with the clarity needed for a new and strange idea. Since a major truncation would almost make a pointless hash of the work, I now sent it, after almost five months of wasted time, to the Royal Astronomical Society.

At his request I had shown my work to Hermann Bondi in March 1948 before it went to the Physical Society, and his interest in the problem of a dynamic but unchanging universe was reawakened. Together with Tommy Gold he tackled the problem from the interestingly different point of view described in Part 1 and, as we have already seen, he eventually managed to inflame most observers against the theory. Now I must confess that both Gold and I contributed in different ways to the prejudice that developed against the theory.

In 1950, Peter Laslett, whom I knew through his Research Fellowship at St. John's, asked me if I would give a series of five forty-minute talks on the BBC "Third Programme." Somebody originally scheduled to give the talks during February of that year had backed out, and there was apparently nobody else who would stand in at short notice.

The talks were a considerable success, thanks to the fury with which Laslett exorcised obscure paragraphs, sentences, phrases, and words from my scripts. The talks were given live at Broadcasting House in London on successive Saturday evenings. We lived entirely from hand to mouth, finishing one talk before thinking at all about the next. Beginning a script on Sunday morning, I would have a rough text ready by Tuesday. Laslett and I would then modify and improve it as best we could until Friday, when Peter's wife, Jan, would type our amended version. On Saturday morning I read through the thing aloud several times, to practice getting the emphasis right, and then it was time to drive once again to Broadcasting House.

I am sure a part of the success of the talks lay in this method of working. Long preparation at a low level of intensity would not have produced as good a result. On the debit side, however, I had gone far beyond anything the academic profession was prepared to tolerate in those days. Undoubtedly, the effect was to impede the acceptance of any of my scientific work by several years. Inevitably, I had made something of a point of the steady state theory, which now began to attract unfavorable comments itself. Bondi and Gold received some of the opprobrium, and in this they were unfortunate, but Tommy Gold was soon to have his revenge on me.

In April 1951, at a scientific conference held in University College, London, Gold had a sharp argument with Martin Ryle which developed later into something of a feud. I have always believed it was this disagreement that first led Ryle to seek a disproof of the steady state theory. When Gold left Britain for Harvard University in

1957, I unfortunately inherited the feud, with consequences that have become all too widely known.

REFERENCES

ALPHER, R. A., AND R. C. HERMAN, 1948. *Nature 162*, 774.

ALPHER, R. A., J. W. FOLLIN, AND R. C. HERMAN, 1953. *Phys. Rev. 92*, 1347.

AMBARTZUMIAN, V. A., 1958. *Report of XIth Solvay Conference*, 241.

BONDI, H., AND T. GOLD, 1948. *Mon. Not. R.A.S. 108*, 252.

BURBIDGE, G. R., AND J. V. NARLIKAR, 1976. *Ap. J. 205*, 329.

CARR, B. J., AND M. J. REES, 1979. *Nature 278*, 605.

FAULKNER, J., F. HOYLE, AND J. V. NARLIKAR, 1964. *Ap. J. 140*, 1100.

FRIEDMANN, A., 1922. *Z. Phys. 10*, 377.

FRIEDMANN, A., 1924. *Z. Phys. 21*, 326.

GAMOW, G., 1946. *Phys. Rev. 70*, 527.

HOYLE, F., 1948. *Mon. Not. R.A.S. 108*, 372.

HOYLE, F., AND J. V. NARLIKAR, 1961. *Mon. Not. R.A.S. 123*, 133.

HOYLE, F., AND J. V. NARLIKAR, 1966. *Proc. Roy. Soc. A 290*, 177.

HOYLE, F., AND J. V. NARLIKAR, 1974. *Action at a Distance in Physics and Cosmology*, San Francisco, W. H. Freeman.

HOYLE, F., AND R. J. TAYLER, 1964. *Nature 203*, 1108.

KRISTIAN, J., A. R. SANDAGE, AND J. A. WESTPHAL, 1978. *Ap.J. 221*, 383.

MCKELLAR, A., 1940. *Publ. Astr. Soc. Pac. 52*, 187.

NARLIKAR, J. V., AND K. M. V. APPARAO, 1975. *Astr. Sp. Sci. 35*, 321.

PEEBLES, P. J. E., 1965. *Ap. J. 142*, 1317.

PENZIAS, A. A., AND R. W. WILSON, 1965. *Ap. J. 142*, 419.

PHLUG, H. D., AND H. JAESCHKE-BOYER, 1979. *Nature 280*, 483.

RANA, N. C., 1980. *Astr. Sp. Sci.*, forthcoming.

RYLE, M., 1955. *Observatory 75*, 137.

RYLE, M., AND R. W. CLARK, 1961. *Mon. Not. R.A.S. 122*, 349.

SEARS, G. W., 1955. *Acta Met. 3*, 361.

SHAKESHAFT, J. R., M. RYLE, J. E. BALDWIN, B. ELSMORE, AND J. H. THOMPSON, 1955. *Mem. R.A.S. 67*, 97.

WAGONER, R. V., 1973. *Ap. J. 179*, 343.

WAGONER, R. V., W. A. FOWLER, AND F. HOYLE, 1967. *Ap. J. 148*, 3.

WALL, J. V., AND D. J. COOKE, 1975. *Mon. Not. R.A.S. 171*, 9.

WILLS, D., AND C. R. LYNDS, 1978. *Ap. J. Supp. Series 36*, 317.

WHEELER, J. A., AND R. P. FEYNMAN, 1945. *Rev. Mod. Phys. 21*, 424.

WOODY, D. P., AND P. L. RICHARDS, 1979. *Phys. Rev. Lett. 42*, 925.

APPENDIX

For a parabolic condition similar to that described in the main text, but for a white hole mass $M_\odot$ instead of $3 \cdot 10^{11} M_\odot$, the following are the formulas corresponding to (7), (8), and (9):

$$R = 1.78 \times 10^{17} \left(\frac{M}{3 \cdot 10^{11} M_\odot} \right) \text{ cm,} \tag{A1}$$

$$\rho_R = 2.54 \times 10^{-8} \left(\frac{3 \cdot 10^{11} M_\odot}{M} \right)^2 \text{ g per cm}^3, \tag{A2}$$

$$T_R = 7.41 \times 10^6 \left(\frac{3 \cdot 10^{11} M_\odot}{M} \right)^{1/2} K. \tag{A3}$$

Since radiation uncouples from matter at about 4000 K, just as before, radiation escapes into intergalactic space at a value of the scale factor S that is changed from the former value of 2000 by $(3 \cdot 10^{11} M_\odot / M)^{1/2}$. This change has the effect of increasing the supply of radiation made available from white holes to the microwave background by $(M/3 \cdot 10^{11} M_\odot)^{1/2}$.

The hierarchical picture described briefly in the text suggests a largest-scale white hole mass comparable to the masses of clusters of galaxies, say $10^{14} M_\odot$, giving $\sqrt{300}$ times the amount of radiation for the microwave background. The continuing expansion of the universe makes some such excess essential. Thus, if the excess were generated at an epoch two-thirds of a generation time ago, the energy density of radiation in intergalactic space would by now have been reduced by exp-8/3, to about $5 \cdot 10^{-13}$ erg per cm^3.

A similar effect of the expansion of the universe also sharply reduces the intergalactic absorptive effect of graphite grains. If the graphite were produced a time H^{-1} ago during a universal epoch of galaxy formation, the opacity per unit path length of intergalactic space would by now have been reduced by exp-3, so that a highly opaque situation could develop to a largely transparent one.

These remarks all relate to the broad form of the steady state theory.

2

STEADY STATE
ORIGINS: COMMENTS I

Hermann Bondi

One's recollections of any event are very much colored by what followed, and so when I write my versions of the origins of steady state cosmology, I am very cautious on a number of points. I can dispense with such caution, however, in saying that there is no question in anybody's recollection that the fundamental idea came from Tommy Gold.

In setting the scene for our discussions, however, I think Fred Hoyle's description needs elaboration. He refers to spending a good deal of time in my room in Trinity. That, one could say, is the understatement of the year. When I got married in November 1947, my wife and I moved to a college flat just outside Trinity at the corner of Trinity Street and Trinity Lane, and Fred continued to spend his days there. In the spring of 1949, a year and a half later, our first child was about to be born and we bought a house a mile and a half from the center of Cambridge. This move terminated Fred's presence in my abode.

Another important factor in the history of the steady state cosmology, which Fred left out entirely and which Tommy reminds me that I must mention, is rum. Rum is an essential part of the history of astronomy, which is so closely connected with the Royal Navy and its rum ration. In the years just after the war booze was very difficult to get hold of in Britain and very expensive. I had an aunt who was an art dealer with a gallery just off Bond Street in central London. The Bond Street traders are rather a club and when one of them, a rum trader, asked my aunt whether she would like a case of rum, she, who had no particular interest in the rum, kindly recalled her very thirsty nephew in Cambridge. My popularity shot up. The inhabitants of my rooms became more numerous and very satisfied. They included—and I think his name should be mentioned here—one of the astronomers who was somewhat older than we, Bill McCrae. He was quite instrumental in obtaining an early hearing and quick publication for the steady state theory.

Neither Tommy nor I can trace the connection between the film that Fred Hoyle mentions and the origin of the steady state idea, though we both agree that the film impressed us very much and Fred may therefore have a point. I do wonder about Fred's time scale, however; he stresses 1946 whereas my feeling is that we really got going on the idea only well into 1947.

Two or three factors were, I think, important for the way the theory developed. One was that Tommy and I were much more philosophical in our outlook than Fred. In particular, we were very impressed by Mach's principle, which did not interest Fred at all at that stage. We were both much more interested in the overall phenomenology of the subject than in the mathematical description, since the latter could not add anything to what one could actually say from a physical point of view. This was a real difference, and it became stronger as the years passed. I think we have always thought it a strength of steady state that it is so rigid as a theory and therefore so easily disprovable. We both expressed this view frequently, whereas Fred never regarded the theory as nearly so rigid. What he discusses today is again rather different; he never has adhered to the more philosophical views of Tommy and myself.

Returning to the history of the matter, I recall that we thought Tommy's first idea was crazy and were sure that we could shoot it down in next to no time. Well, we couldn't. Instead, we gradually became rather convinced of it. There is no doubt that in those days all cosmological thinking was heavily burdened by the time scale difficulties. Nonetheless, we felt extremely cautious about publishing, because steady state was only an idea and, although it dealt with the problem of time scale, we did not think that simply publishing an idea would be wise—if indeed it were publishable at all. Fred then began his C-field computations, with no encouragement at all from us, very much on his own. We were both a little bothered (and I think this is an occasion for frankness) that something that had originated with Tommy and on which we all had worked was likely to be made public on the basis of what we regarded as uninteresting computations. As an aside, I might add that it seemed rather funny that it was *he* who went into mathematics. Of the three of us, I was the only one who could handle mathematics deftly. The physics, the intuition, belonged much more to Tommy and Fred, but here was Fred going off with what appeared to be a piece of mathematics serving one particular purpose. We were somewhat disturbed that we had no peg to hang our coat on, as it were, no point that could serve as the basis for publication.

One evening late in 1947, while looking for the umpteenth time at Hubble's number counts (which by now of course, are totally discredited), I noticed that if one adjusted the only free parameter, the Hubble constant, those counts seemed to fit steady state very well. I was so excited that I rushed down (we had no telephone in our flat) to the little phone box in Rose Crescent and phoned Tommy. He immediately said, "That's it!"—and thus we wrote the paper. It is interesting that our paper was stimulated by the Hubble counts of galaxies to which nobody would give any value now. It was then due largely to McCrae, who was the secretary of the Royal Astronomical Society, that both our paper and Fred's paper got very rapid publication in 1948. All three of us talked at the meeting of the Royal Astronomical Society in Edinburgh in November 1948, which is how the theory received its first public presentation. From the beginning, however, McCrae was inclined very favorably to the idea. In the summer of 1948, when the International Astronomical Union met in Zurich, he strongly encouraged me to talk about it to various of the scientists who were present, including Bertil Lindblad.

I did not remain active in the field for long—the last paper I wrote on cosmology was a joint study on the radio number counts, published in 1955—but from early on I kept challenging adherents of evolving models. I told them that if the universe had ever been in a state very different from what it is today, they should please show me some fossils of that earlier age. At that time, there was no answer at all to this challenge. I began to suspect that the amount of helium might be very important as a potential fossil and talked about it to Andrew McKellar. I think it was at the same time that Edwin Salpeter showed that with a hot big-bang only helium and no heavier elements would be left. I must confess that the three degree radiation did not cross my mind.

I might finish with one other bit of history. Tommy has now been connected with Cornell University for a long time but the first connection of any of the three of us with Cornell bears mentioning. Since I had been lecturing in Cambridge since 1945 and thought I deserved a sabbatical term for the spring of 1951, I wrote to various eminent American institutions, including Cornell. I can't remember what happened with all of them. One sent me a form, another wasn't particularly thrilled, but I received a letter almost by return post from Hans Bethe, who said how glad he would be if I came. He was sure that it could all be arranged, and added that, since I had a wife and young child and would presumably want to find my own accommodation, I should stay as a guest in his house while looking. Such warmth made it very easy for me to decide where I should go. This was the first link that any of the three of us had with this great university.

3

STEADY STATE ORIGINS: COMMENTS II

Thomas Gold

In a few words I would like to discuss the old relationship of the three of us, and not just as it concerns the steady state, for Hoyle, Bondi, and myself have done other work together as members of various groups. There is no question that I was the junior partner both in age and in terms of knowledge and experience and insight into science. Little of my scientific education comes from my Cambridge undergraduate career. Being interned in rather miserable circumstances as a supposed enemy alien for one year of the war with Hermann Bondi was probably more important in getting me started on a scientific career. As there wasn't much else to do, he taught a number of us about subjects that he, at that stage, was extremely expert in, like dynamics. I remember being given problems of vector analysis for calculating a rolling sphere on a cylinder and other similar exercises. But he also talked to me quite a lot about mathematical physics, and especially about the outlook he always had—an outlook that went quickly to the heart of a problem and placed it in the proper context.

But there is also the relationship between Fred Hoyle and us to be discussed. During the war we worked together and also roomed together for some time. In the evening Fred would typically walk around and with great emphasis say: "Well what could that Hubble observation mean? Find out what it could mean!" He would continue along this line, sometimes being rather repetitious, even aggravating, drumming away at particular points without any obvious purpose. At other times Fred would have Bondi sit cross-legged on the floor, then sit behind him in an armchair and kick him every five minutes to make him scribble faster, just as you might whip a horse. He would sit there and say: "Now come on, do this, do that," and Bondi would calculate at furious speeds, though *what* he was calculating was not always clear to him—as on the occasion when he asked Fred, "Now, at this point do I multiply or divide by 10^{46}?"

As to the origins of steady state, I think my version is similar to that of Hermann Bondi. When I came up with this outrageous idea and presented it to the two of them, I at first got a rather brief reception: "Well, by tomorrow we'll have shot it down." They advanced a number of objections, but I countered them in one way or another. For example, in response to their point about conservation of energy, I said: "Well, look, if you make it in one big bang to begin with, what are you doing then? You are not conserving energy then either, so somewhere you've got to do it, whether you do it steadily with a definite law, or with some single event, I couldn't really say which is better; if anything, I prefer the steady one." Furthermore, I had understood by that time that in a universe of a particular density you can really not say that the energy is being increased by inserting a new particle, because the binding energy of that particle in the gravitational field of the rest can just balance its rest mass. And so various complex arguments were produced to show that my idea was not nearly as objectionable as it had seemed at first. Fred is right when he says it wasn't long before we understood that we had to create matter. It wasn't long at all, because when I first proposed the idea I immediately added that of course you had to create matter and let it drift apart, just arranging that the speed of condensation into galaxies matches the speed at which they move apart. I had even gone one step farther, saying that if the speeds did not match then the matter would simply do as it liked until it reached the point at which the condition of density and the expansion speed did match. Many physical problems share this quality: you can start them out wide of the mark and yet they will find their steady state solution. So all such questions were settled right away, of course.

Hermann slowly warmed up to the idea and said: "Well, you know, it isn't so crazy after all and has to be taken seriously. I thought I would be able to shoot it down right away, but as it is, it looks more and more interesting." He was spurred on not only by the number count story that he recounted in these pages, but also by the fact that he could calculate that a particular metric was singled out as the only one fitting, namely the de Sitter metric. This geometry of the universe then became very much more meaningful than it was in de Sitter's original discussion.

I also recall an event that Ray Lyttleton unfortunately does not recall, but it is so firmly fixed in my mind that I can't be wrong about it, and that is that Fred walked along, in fact through Caius College—I know exactly where—with Ray and me, and Ray said to Fred something in the nature of "What kind of cosmology do you think is reasonable?" Fred replied, and that was the first indication I had of his change of heart in the matter, "Well, really I think the only chance is Tommy's proposal of the steady state." And, I suppose it was not very long afterward that he came to me, rather abruptly, and said, "I have written a paper, almost completed it, and I would like you to allow me to put your name on it as well, as a joint publication." At that stage I said, "Well, let me look at the paper, but I really am already beginning to write a paper with Hermann and we have a rather different philosophical outlook on the matter. I may not wish to have my name on your paper." Although he had written the whole paper, I suppose he very properly and correctly thought that since the idea had come from me, a joint publication was appropriate. When I looked at the paper it was, for my taste, much too specific. I thought that this idea should be presented in its complete generality rather than tied to one particular mathematical construct that was not absolutely necessary.

At this point, I began to spur Hermann to proceed with our paper, after which we wrote rather quickly. The paper contains what we thought to be weighty general arguments—arguments that we had thrashed around many times before and that now needed only to be put down on paper. We finished the paper quite easily and then, by a curious accident, Fred's already submitted paper was delayed in the Royal Astronomical Society by a referee, and our paper thus appeared first. The text of our paper already refers to his paper, however, so it is clear that we had all seen each other's work. That, I think, is the best recollection I have of that part of the story.

As to the apportioning of credit or discredit, I was, as I have said, undoubtedly the junior partner. My interest was in reading Jeans' and Eddington's semipopular books, and so on, but not much beyond that. I was working in other branches of physics at the time, and it was certainly Hoyle who produced a great deal of the vigorous conversation. As far as I was concerned, the teaching that I received from Bondi was the important thing; indeed, I would have been a complete dilettante at the business without him. I think I produced a number of good ideas, not just this one but in several other instances. But over the years, I became more knowledgeable in many areas. That whole attitude of, "you know, you really have to learn things and really have to know things," was very much Hermann's attitude. I would have been a playboy in science, I think, without that.

4

THEORY OF GRAVITATION

Hermann Bondi

Einstein's theory of gravitation, misnamed by him (in my view) General Relativity, has a bad name for being obscure physically and absurdly complex mathematically. I want to show here that far from having given us an unnecessarily difficult theory, Einstein has in fact given us the simplest possible theory compatible with Galileo's principle that all bodies fall equally fast. While I cannot abolish the mathematical complexity, I aim to show that the physics of the theory is as transparent as Galileo's discovery permits.

To say that every body behaves in the same way sounds like the simplest of all possible ideas. Simple perhaps, but most certainly not harmless. As an analogy, suppose you lived in a world much simpler than ours, a world so simple that all bodies had the same coefficient of thermal expansion—very nice indeed, until somebody asks you to construct a thermometer! This is precisely our difficulty with gravitation, for the absence of a nonfalling standard is the crux of the problem.

It is always much easier to use hindsight than foresight. For us today, much that was difficult to understand, not only for Galileo's contemporaries but also for people who lived much later, has been made obvious by the progress of technology. The notion of weightlessness in free fall has been talked about for a long time but only for two decades or so have we been able to observe astronauts in free fall in their orbiting spacecraft. When they are falling freely, as we all know, gravitation seems to have disappeared. Drops of soup float around the room; indeed, everything floats, gravitation appears to be abolished totally, and so one wonders what we are really talking about when we use the term. From the day that Newton, basing himself on Galileo, worked out his dynamics and his theory of gravitation, it was immediately thought to be a *universal* theory, valid everywhere and anywhere. Free fall is not our common lot on the surface of the earth, due to the particular strength of the material of the earth, but in the universe, it is rare to find oneself on the surface of a solid body. Falling freely, like a spacecraft, is something that many more particles and bodies experience. Is gravitation, therefore, a figment of the imagination, a local prejudice that springs from living on the solid earth?

It requires closer analysis to show that gravitation cannot really be abolished. Consider Figure 4.1, in which a spacecraft, B, orbits about the earth, A. As you know, the spacecraft is falling freely and everything in it is falling freely, apparently abolishing gravitation. The spacecraft is not a point, however; it is a body of a certain size. The part of the spacecraft nearest the earth will want to fall a little faster than the middle of the spacecraft, which will want to fall a little faster than the part farthest from the earth. Now, one could say that if your spacecraft was made by a reputable manufacturer, this stress is nothing to worry about. But consider a particle of dust X close to the

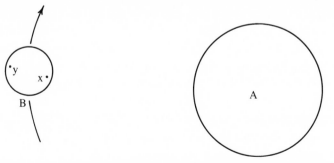

Figure 4.1. Dust particles in a spacecraft (B) orbiting the earth (A).

part of the spacecraft nearest the earth. The spacecraft, as a whole, will fall with some compromise acceleration. The little bit of dust X will want to fall a little faster and so will gradually settle on the part of the spacecraft nearest the earth. Similarly, a particle of dust Y near the part of the spacecraft farthest from the earth will want to fall a little more slowly than the spacecraft as a whole, and so will drift toward the part of the spacecraft farthest from the earth. Thus, whenever you clean a spacecraft, always remember to clean particularly well the part nearest the earth and the part farthest from the earth. The dust-conscious astronaut, though gravitation seems to have been abolished for him, will notice the way the dust accumulates and, in so doing, observe something about the gravitational pull of the earth in spite of his falling freely.

This is a residual effect which is a universal *observable* and thus, in our modern view, it is the essence of gravitation. Though all bodies in the same place fall equally fast, particles that are a little apart from each other will fall a little differently. Interpret the diagram in a slightly different way. Think of A not as the earth but as the Sun and of B not as a spacecraft but as the earth. The earth, after all, is freely falling in its orbit about the sun. As a result, we will observe a force to elongate the earth along the line linking it to the sun, both on the side nearest the sun and on the side farthest from the sun. The most easily observable effect of this force will be on the very visible part of the earth which most readily follows this pull—the ocean. We call this effect the solar tide. The lunar tide, which is rather more than twice as large, is caused in the same manner, but the orbit and radii of the bodies are in very different proportions. The solar tide is easily observable and is a direct evidence of the existence of the sun and the orbital motion that we follow, just as the lunar tide is a direct evidence of the existence of the moon and its orbit about the earth.

I sometimes like to imagine a civilization without space capability living on a wholly cloud-covered earth and deducing the existence of the moon and of the sun from the observations of the tides. It would be very interesting to elaborate exactly what you could discover about these bodies through the observation of the tides, what you would learn about them without ever looking at them.

I have said, therefore, that given Galileo's dictum that everything falls equally fast, the one universal observable of gravitation is the differential acceleration of neighboring particles. How can one describe this differential acceleration mathematically? It is a differential acceleration that vanishes when the particles are coincident. It is thus appropriate to assume a *linear* relation between the separation vector of the two particles and the vector representing their differential acceleration, the differential acceleration that shows itself in the tides and in the dust accumulations in the spacecraft. The observable of gravitation is therefore the entity that links linearly the differential acceleration vector, ∂f^i, with the displacement vector, ∂x^j. As is well known such a link is the tensor, a^i_j in the relation (1) $\partial f^i = a^i_j \partial x^j$. Hence the universal observable of gravitation is this tensor, which must be symmetrical, for otherwise it could speed up the spin of a sphere and generate energy without any other change, a situation that is not permissible. In Newtonian theory, in which gravitational acceleration is the first derivative of a scalar, this tensor is thus composed of the second derivatives of a scalar, the gravitational potential:

$$a_{ij} = - \frac{\partial^2 V}{\partial x^i \partial x^j}.$$ (1)

In this presentation the only observable is this tensor, and therefore neither the magnitude of the potential nor its gradient are observables, only the second derivatives being such. The other half of Newton's theory of gravitation is that the link between gravitation and its source, which Newton recognized as mass, is given by Poisson's equation that $\nabla^2 V$, which is a certain linear combination of some of these second derivatives, is proportional to the density of matter ρ, so that

$$\nabla^2 V = 4\pi G\rho,$$ (2)

G being the constant of gravitation. This completes the presentation of Newton's theory in modern form.

Newton's theory of gravitation was perhaps *the* best established theory that there has been in physics. For well over two hundred years it was outstandingly successful in predicting the positions of the planets and of their satellites, in not only forecasting eclipses of the sun and of the moon, but also in hindcasting them and so being able to date historical events. It seemed to be the very prototype of a physical theory.

Perhaps one of the theory's greatest successes was the way it led to the discovery of Neptune, the history and nature of which has been so well described by Raymond Lyttleton. The discovery was a fine feat of using the theory to account for small observed discrepancies in the motion of Uranus to indicate the body that had to be responsible for them. But another discrepancy was found, one relating to the motion of the planet closest to the sun, Mercury. Just as the deviations in Uranus were used to work out where Neptune should be, so one used the minute but gradually well established deviation of the motion of Mercury to work out the motion of a hypothetical planet still closer to the sun than Mercury, a planet that was named Vulcan. It was thought that it was impossible to see Vulcan because its close proximity to the sun meant that it was obscured by the sun's glare. I believe work on Vulcan started about one hundred fifty years ago, and only in this century did better observing equipment allow one to see nearer to the sun, thereby raising severe doubts about the planet's existence. Though the importance of this blow to the credibility of Newton's theory was not at first appreciated, we can now see that his theory has other weaknesses, just two of which I wish to discuss here.

Newton's theory of gravitation, like Galileo's principle earlier, was a statement about *material bodies*—material bodies like the planets and their satellites. The relation of gravitation to light is a different question, and indeed a logical synthesis of the theory of propagation of light with Newtonian gravitation becomes appallingly difficult. There are essentially two options: either light is unaffected by gravitation or light moves like a particle moving with the speed of light.

There is nothing in Newtonian theory to rule out the notion of particles moving as fast as light or even faster. The idea that such fast particles are bent in their orbit around the sun while light goes straight is very disconcerting. But it is even worse to assume that light moves like a particle, because then the speed of light would necessarily be greater near the sun than far from the sun. The speed of light would indeed be determined by the potential. Any apparatus that could measure the speed of light would therefore determine the gravitational potential—a situation quite contrary to the previously stressed idea that both the potential and its first derivatives are not observable, with only the second derivatives observable. And if anything makes a mathematician shudder, it is something that makes invariance properties go awry. This difficulty with light is simply a demonstration of the fact that *Newtonian gravitation is not a relativistic theory*.

Still more awkward is the question of the energy of gravitation. Every conventional textbook states that the potential energy of gravitation is negative. When first told about this, almost everybody feels uncomfortable because all other energies are positive. Kinetic energy, $\frac{1}{2}mv^2$ is obviously positive definite. The energy of an electrostatic field is $E^2/8\pi$, which is necessarily positive; the energy of a stretched spring is necessarily positive. How can it be, then, that among all these necessarily positive energies we suddenly find a negative one? The textbook answer is that only differences of energy are significant, and that the absolute value therefore matters little. Indeed, the textbooks suggest that with gravitation we *happen* to have chosen the fully dispersed state of matter as the zero. Therefore, when matter condenses under the pull of its gravitation, it generates energy. When this energy is extracted, something less than zero, something negative is left. The problem, however, is not that we *happen* to have chosen the fully dispersed state of matter as the zero; it is much more deep-rooted. Consider a contracting star. As it contracts further, it will necessarily generate more energy. Each further contraction will yield further energy, though of course this energy often will be absorbed by the compression of matter. There is nothing in the theory of gravitation that says that this process should come to an end. However condensed a body is, it can gain energy from gravitation by contracting further. In the case of the earth, only its elastic properties prevent it from condensing further. If we could leave these properties out because they are not part of the theory of gravitation, there would be no energy problem; whenever we were a little short of energy we would allow the earth to contract. The remarkable point is that if the earth contracts, gravity goes up; further contraction becomes more profitable, and this process would continue without limit. The negativeness of gravitational energy is thus only a sympton, the real problem being that *gravitational energy is not bounded from below.* Indeed, the more energy is removed, the stronger gravitation becomes—a situation in flagrant contrast with all other kinds of energy. Think of kinetic energy, the energy of motion. If the brakes are put on, the motion diminishes as does its energy, and once the system has been brought to a stop, the phenomenon of motion is gone and the kinetic energy is zero. Similarly, the energy of the stretched spring, due to the tension in the spring, is the energy that was put into extending the spring. It can do work, that is, lose energy, if allowed to contract. This contraction will diminish the phenomenon, the tension of the spring, and when it reaches its natural length, all the energy and all the tension are gone. Only in gravitation does the removal of energy increase the phenomenon, increase the force.

The potentially limitless source of energy which gravitation produces makes us uncomfortable, because we always regard energy as a commodity of which only a limited amount is available. Since one cannot create arbitrary amounts of energy, energy equations are awfully useful in physics. In complex dynamics (e.g., the motion of a bicycle) the energy equation will be very important. But if (and I leave out the elasticity of steel) any ball in the ball bearing of a bicycle could generate as much energy as it liked from self-gravitation simply by contracting sufficiently, the energy equation would be useless, just as money would be useless in an economy where everyone was allowed to print his own. Here there is a real problem.

How, then, can we make a relativistic theory of gravitation, thereby dealing with the problem of light propagation and additionally resolving the energy paradox of Newtonian theory? Such a theory might best be evolved from the consideration of a situation in which gravitation and the propagation of light are both involved. One such example, concerning the interaction of light and gravitation, was first presented by Einstein, much improved by Tommy Gold, and often used by me. Consider a tower with a chain carrying buckets which runs over wheels at the bottom and at the top of the tower (see Figure 4.2). Fill each bucket on the chain with the same number of atoms of the same element, taking care to ensure that all the atoms on the side labeled E are in a certain excited state, while all the atoms on the side

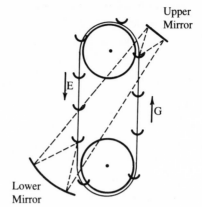

Figure 4.2. The tower experiment.

labeled G are in the ground state. As is well known, an atom in the excited state has more energy than an atom in the ground state. It can get rid of this excess energy by radiating light of just the right frequency so that, if absorbed, it puts an atom that is in the ground state in the excited state. As special relativity shows (and countless experiments have confirmed) the atoms in the excited state, having more energy than the atoms in the ground state, will also have more mass. As Galileo taught us (and as L. Eötvös and R. H. Dicke have since confirmed to great accuracy), more mass means more weight. Side E therefore will be heavier than side G and the chain will begin to move. As is allowed by physics, we tickle the excited atoms when they come to the bottom, so that they make the transition to the ground state, emitting light that is caught by a big mirror and sent to the top of the tower. This light in turn irradiates the atoms as they arrive from the bottom. Since the transition from the excited to the ground state produces light of the frequency required to put atoms from the ground state into the excited state, the system goes on working. Side E will continue to be heavier than side G, and so the chain will go on moving. Indeed, its continued motion can generate useful energy by, say, driving a generator from one of the wheels. Thus the system should create energy, contrary to the principle of conservation of energy.

Thus, the analysis just given must contain a flaw invalidating it. Where can it have gone wrong? Not in assuming that there exist both

excited atoms and atoms in the ground state, for we know that this is
so. Not in saying that excited atoms have more energy than atoms in
the ground state, as we know this to be the case. Not in the sense that
more energy means more mass, because the theory is quite clear on
this point, and although it has not been possible to test it experimen-
tally in the atomic situation, it has certainly been established in the
nuclear situation. Not in the sense that when we have excited and
ground state atoms side by side, in suitable circumstances, the de-
excitation of the one can excite the other, as this fact is also well
established. The only thing that can be wrong is that if the de-
excitation is at the foot and the atoms in the ground state needing to
be excited are at the top, then, although the frequency was correct
down below, it is no longer correct at the top. What shift of frequency
could make it impossible for the light to excite the atoms arriving at
the top? Only if the light has insufficient energy, if it is too red. In
order to get the frequency right one would have to blue shift the light,
that is, to reflect it from an advancing mirror. We therefore put a
series of mirrors on a wheel and spin it (see Figure 4.3). If the light
had been too red, things will now work as they were supposed to
provided the wheel is turning at the right speed. Energy is still being
generated by the chain but is now needed to turn the wheel of mirrors,
because the wheel turns against the pressure of the incident light.
Energy conservation makes it clear that the chain produces just
enough energy to turn the wheel of mirrors. Thus, the ideal experiment establishes that there is a *gravitational red shift*, that light
produced at the bottom of the tower will look redder when viewed
from the top than it did at the bottom.

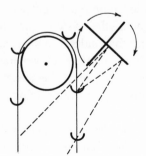

Figure 4.3. The top of the tower with
the rotating set of mirrors.

People tried hard to test this exciting result of Einstein's by looking at light from the sun and comparing its spectral lines with similar ones produced in the laboratory. Unfortunately, there are other complex factors that are not easy to calculate and that tend to hide this minute gravitational shift because it is not possible to reproduce accurately in the laboratory (where one doesn't have available a few tens of thousands of kilometers) the process by which spectral lines are produced in the sun's atmosphere. But when the Mössbauer effect came along with its incredibly sharp gamma ray lines, R. V. Pound and G. A. Rebka (in a celebrated experiment conducted twenty years ago) actually measured the gravitational red shift between the foot of a tower and its top. Some people were rather worried about an apparent conflict between theory and experiment during those agonizing years of unsuccessful astronomical verification; Gold and I were quite unconcerned. A sound logical deduction from theories that rest on accurate, well-established experiments must carry more weight than experiments at the extreme limit of what is possible and are not easy to interpret. Nonetheless, even we found it delightful when Pound and Rebka demonstrated the gravitational red shift.

It sounds very erudite and of small general relevance to have a shift between the foot of a tower and its top which is so small that the best-available experimental means can just barely measure it. But such a shift has very important consequences, not just because spectral lines are the way that spectroscopists make their living (important though this fact no doubt is), but because spectral lines are the basis of timekeeping. Whether one employs a complicated, atomic or molecular clock that uses ammonia or caesium and gives the most accurate timings now possible or a watch with a balance spring that uses the phonons of steel; whether one uses a quartz-controlled watch or, like the archaeologists, the methods of dating by radio carbon invariably a spectral line (or something dependent on a spectral line) defines the measure of time. The physicist is by nature a skeptic to whom the idea that time exists as an independent concept does not appeal at all. Every quantity has to be defined, as P. W. Bridgeman taught us, by the means of measuring it. Time is that which is measured by clocks or, as I prefer to put it, time is that which is manufactured by clocks. And what we are saying is that the time manufactured by clocks at the bottom of the tower is not the same as the time manufactured by clocks, identical in construction, at the top of the tower. Einstein's great contribution to special relativity in 1905 was to show that time was different for *differently moving* inertial observers. His theory of gravitation showed that time was also different for *differently located* clocks in a gravitational field. And so we go a step further in the disintegration of the time concept, a step that implies a further departure from familiar concepts, on this occasion in geometry.

A space-time diagram (see Figure 4.4) which is drawn as is normal in physics—time is vertical and height horizontal—depicts the tower experiment. The line ABCD represents the foot of the tower and EFGH the top. The inclined lines BF and CG represent two successive light signals traveling from the foot to the top. Since nothing changes in the system, BF and CG are parallel, as are BC and FG, and BFGC is therefore a parallelogram. But the red shift implies that local clocks measure *different* intervals BC, FG, as a result of which this parallelogram has opposite sides that are unequal. But a parallelogram with unequal opposite sides is impossible in Euclidean geometry, and the time-height 2-space is therefore non-Euclidean. At first sight this situation seems unexciting, for we deal all the time with non-Euclidean 2-spaces, such as the surface of the earth. But now recall that the tower stands in a spherically symmetrical system, our earth and its surroundings. As soon as one combines this non-

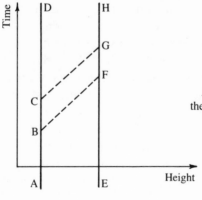

Figure 4.4. Height-time diagram of the tower experiment.

Euclidean 2-space of time and height with the 2-space of longitude and latitude, a non-Euclidean 4-space necessarily results. With spherical symmetry, the time-height 2-space of the tower cannot be embedded in a Euclidean space-time. Thus, we are driven to non-Euclidean geometry, which is not a simple concept. There are various non-Euclidean geometries and, Einstein having been far from malicious, chose Riemannian geometry, the simplest form of non-Euclidean geometry. But the point I want to stress is that it was Galileo who forced us away from a Euclidean to a non-Euclidean space-time. The complexities were inherent in his discovery that all bodies fall equally fast.

The formulation of Newtonian theory given above led to the three-dimensional tensor a_{ij}. Clearly, to make it relativistic means to make it four-dimensional, fitting it into space-time. But this is not all. The Newtonian linkage between the displacement and relative acceleration vectors was independent of velocity, a situation that is not compatible with special relativity, which cannot allow rules that make it possible for a particle to be accelerated from below (even if only just below) the speed of light to above that speed. The law of relative acceleration (as stated above) would make it possible to find speeds of particles and displacements that would lead to this result.

Thus, not only is it necessary to make a_{ij} four-dimensional, it must also be made velocity dependent. The symmetry of a_{ij} and the structure of the four-dimensional velocity vector v^i lead, as the simplest relativistic formulation, to

$$f^i = R^i{}_{jkl} \, V^j V^k \, \partial x^l \qquad (3)$$

where the four-suffix tensor R has quite recondite symmetry properties and now is the observable. For low velocities (v^i close to 1,0,0,0) this tensor reduces to the previous formulation, while for high ones the formula rules out a transgression of the speed of light.

In the Riemannian geometry of space-time, the difference from Euclidian space-time is described by an entity (the curvature tensor) identical in its character with the R tensor just defined. Indeed, this curvature tensor governs the relation between geodesics (the nearest thing to straight lines in a Riemannian geometry) in exactly the manner R governs the differential acceleration between two freely falling particles. We are thus driven to the conclusion that in the Riemannian space-time necessitated by the tower experiment, geodesics represent the paths of freely falling particles, and the gravitational field is represented by the curvature tensor.

The logical appeal is strong. In Newton's dynamics, the concept of force is that of something that may or may not be present, something that can be switched off, whereas inertia is ever present. Thus a magnetic force on a body can be switched off by demagnetizing it, an electrical one by neutralizing it, and so on. But no body can divest itself of its inertia or, indeed, since all bodies fall equally fast, of the effect of gravitation. Inertia and gravitation should thus be viewed as one and the same inalienable property of bodies, and gravitation should not be considered a "force." Therefore, Newton's first law of motion should in logic be reformulated as: "A body on which no force acts moves along a particular type of orbit, the path of free fall, which is represented by a geodesic in space-time." The geometry of such paths (not quite as simple as that of straight lines) is described by the curvature tensor of the space-time. A force, in the sense of Newton's second law, is then something that pulls a body away from such paths.

We have arrived now at a wholly self-consistent and relativistic version of the formulation of gravitation, given in nonrelativistic form in (1) and now written as (3). What about the source of the field, given in nonrelativistic form in Poisson's equation (2)?

It is relativistically sensible to suppose that if mass causes gravitational fields, so do energy, momentum, and stress. (This assumption is not obligatory, but the only real alternative is to consider rest mass as the sole source of the field—an alternative that is physically awkward and has no mathematically agreeable link to the tensor R.) Following (2), in which a linear combination of some of the observables was utilized, we look to a linear combination of our relativistic observables, the R tensor. A certain such combination, the Einstein tensor G_{ij}, has the right symmetry and moreover possesses conservation properties such that putting it proportional to a tensor T_{ij}, which describes the densities of mass, energy, momentum, and stress, immediately implies the conservation laws for these quantities (essentially in the form of Euler's equations of hydrodynamics). Thus we replace Poisson's equation (2) by (choosing suitable units)

$$G_{ij} = -8\pi T_{ij}, \tag{4}$$

known as Einstein's field equations. They are superior to (2) not only in being relativistic but also in necessitating the conservation rules for the sources of the gravitational field. (In Newton's theory, mass was the source of the field but its conservation law was an extraneous constraint apparently unconnected with the nature of gravitation.) This completes the task of making Newton's theory of gravitation relativistic. How does this theory of gravitation fare in the empirical testing?

First, it emerges that for slowly moving bodies the answers obtained by this theory are very close to the Newtonian ones, and all that Newton had achieved in describing motions in the solar system is thus maintained. There do occur small deviations that increase with velocity so that the fastest moving planet, Mercury, will have the biggest and indeed the only practically observable deviation. These deviations have also been discussed for Venus, the Earth, and Mars, but all are on the borderline of attainable observing accuracy. The established deviation of the motion of Mercury is perfectly accounted for; the propagation of light past the sun is also measureable and again agreement between theory and observation is good; and other tests are described elsewhere in this volume. We can therefore conclude that, thanks to Einstein, we now possess a good theory of gravitation.

How does this theory deal with the bottomless pit of energy that Newton's theory suffered from? When you ask what you believe to be a profound question in physics, and eventually your equipment and your theories improve to the extent that you are able to get an answer, it usually turns out to be a lot more sophisticated than could have been envisaged previously. Regarding the question about the negative nature of gravitational energy, a good Einsteinian theory of gravitation cannot deny any more than a Newtonian one that when dispersed matter condenses, energy is released. But what happens in further condensation, from which it seemed that we could extract arbitrarily large amounts of energy? Applying an Einsteinian theory, which is relativistic, we take mass away when we extract energy from this contracting body, because energy has mass. (In the Newtonian picture, the mass stayed constant no matter how much energy we extracted. If enough energy, enough mass, has been taken away, a highly condensed body remains, but it has no mass and therefore no energy. Thus there is a lower limit (admittedly somewhat obscure) of the energy that could be extracted; there is a lower bound to recoverable gravitational energy. It is easily shown that at the most we can extract rather less energy than that which corresponds to the rest mass of the originally dispersed matter. If the energy is not extracted, however, if it stays put, then I can see no limit to the possible fate of the local gravitational energy, but this fact, though uncomfortable, does not constitute a paradox.

I trust I have shown that the problems that arose largely from Galileo's notion that all bodies fall equally fast, problems that became so extreme when we tried to unite this notion with ideas of the propagation of light, are really not all that difficult. Though the mathematical technicalities become pretty awkward, the physics is clear and intelligible. Einstein's theory of gravitation emerges readily once Newton's theory is viewed in a modern way, looking at what is observable, and uniting it with what is known about the propagation of light.

In this presentation there have been two casualties. First, what is often called a "uniform" gravitational field, one in which there is no relative acceleration between neighboring freely falling particles, is, in the language used here, no field at all. Second, there is neither room nor need for any "principle of general relativity," one that has always seemed meaningless to me, as to E. Kretschmann and V. A. Fock. The term "general relativity" will be hard to expunge, but Einstein's theory of gravitation is a superb theory not dependent at all on this irrelevant and vacuous idea of "general relativity." Einstein's theory is not only the best, but the simplest possible theory of gravitation compatible with Galileo's finding that all bodies fall equally fast and with what we know of the propagation of light.

PART II

HIGH ENERGY ASTROPHYSICS

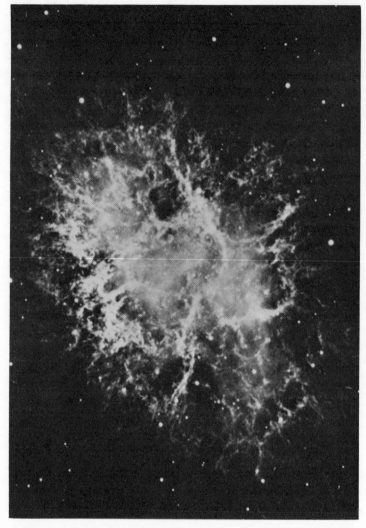

Plate 2. The Crab Nebula, a supernova remnant. Courtesy, Palomar Observatory, California Institute of Technology.

5

BLACK HOLE EXPLOSIONS

Dennis W. Sciama

When I was invited to contribute to this volume I was overjoyed as well as honored. I am one of the many people who received crucial help from Tommy Gold at the beginning of their research careers. Such help is a precious gift that one cannot possibly repay. All one can really do is to express gratitude, and where better than here?

Tommy helped me primarily by emphasizing two points of view toward physics. First, he stressed its essential unity. There should be no plasma physics or solid state physics or relativity but just physics, with thermodynamics acting as an overall constraint on the subject (of which more anon). This attitude, of course, is widely held among good physicists and not unique to Tommy, but his second major influence is perhaps more peculiar to him. It involves the fact that, while the main basis of physics is experimental, the results of experiment become codified into an elaborate and powerful formalism. This formalism comes to dominate the subject, and it is proper that this should be so. But Tommy taught us that sometimes this domination is too strong, that in an important sense the true essence of physics is not quite encapsulated by the powerful formalism but lies slightly elsewhere. It is to the study of this "slightly elsewhere" that

Tommy has devoted his working life and, accordingly, has influenced all those disposed to listen to him.

In thinking over the choice of topic for my paper I looked for one that illustrated these two points of view and also related directly to one of Tommy's own finest discoveries. I did not have to look far. For some time I have been trying to demystify Stephen Hawking's remarkable result that black holes, far from being the most passive objects in the universe, are actually the most active, radiating like hot bodies and ending their days in the most violent explosion known to man. Now most physicists have taken black holes seriously only since Tommy showed conclusively in 1968–1969 that the pulsars that had just then been discovered must contain rotating neutron stars. His arguments were rapidly and widely accepted by the community of physicists and astronomers, and, as we shall see, once it is clear that neutron stars exist in the galaxy it is a short step to accepting that black holes exist there also.

I shall try in this paper to give a simple physical account of black hole explosions, avoiding the heavy formalism with which the literature on this subject abounds. But how about Tommy's other hobbyhorse, the unity of physics; where does that come in? The answer to this question is simple. Indeed I know no other process in physics whose very nature arises from a unification of so many different branches of physics. Jacob Bekenstein foreshadowed and Hawking established that one can attribute a temperature and an entropy to a black hole in such a way that Einstein's field equations imply that all the laws of thermodynamics are satisfied by processes involving black holes. Since the arguments involved are essentially quantum mechanical, one has thereby brought about a deep unification of general relativity, quantum field theory, and thermodynamics.

We may recall that whenever two different branches of physics are unified, both the conceptual and the practical consequences are usually immense. One thinks of Faraday uniting electricity and magnetism, Maxwell uniting electromagnetism and optics, Einstein uniting mass and energy and then inertia and gravitation, and Dirac uniting special relativity and quantum mechanics. Here we have three branches of physics united, with consequences that we are just beginning to explore.

Neutron Stars

Neutron stars represent an extreme form of matter in which approximately a solar mass of material is squeezed into a radius of about 10 kilometers. The resulting object has a density of about 10^{15} grams per cubic centimeter, the same as in the nucleus of an atom. Indeed the whole neutron star can be considered as a gigantic nucleus with an atomic weight of 10^{57}! The main difference is that the short-range attractive nuclear force that holds a real atomic nucleus together is here swamped by the gravitational force of the solar mass object. This gravitational force is very strong and one needs to find a form of pressure gradient capable of balancing it and so preventing the neutron star from collapsing under its own gravity. An ordinary star is supported by thermal pressure, but in a neutron star all heat sources have burnt themselves out and adequate thermal pressure is not available.

At this point we make our first appeal to quantum mechanics. Neutrons obey the Pauli Exclusion Principle, which forbids two fermions from occupying the same state in a given physical system. Thus, even if a neutron star were quite cold, its constituent particles would necessarily occupy a whole range of energy states in which they would carry momentum from place to place in the star. The ideas of the kinetic theory of gases may then be applied and one finds that a pressure gradient is present, although one must speak of degeneracy pressure rather than thermal pressure. This degeneracy pressure is fundamental in ordinary physics, since it is the agency that prevents a solid body from collapsing under the electrostatic attraction of its constituent protons and electrons. In that case, most of the degeneracy pressure is exerted by the electrons, because of their low mass. The same is true in a white dwarf, a solar mass object of density about 10^6 grams per cubic centimeter in which, as in a neutron star, the attractive force holding the object together is mainly gravitational rather than electrostatic.

It is important to understand what happens when an object like a white dwarf or a neutron star of greater and greater mass is considered. To overcome the increased gravity one needs a greater degeneracy pressure. This pressure can be obtained by reducing the size of the object, since by forcing its particles closer together in space they are kept farther apart in energy and momentum in order to satisfy the Exclusion Principle. This increase in momentum then leads to an increase in the degeneracy pressure that can be exerted, and the increased gravity can still be balanced. One finds in this way that the mass M and the radius R of the object are related by

$$MR^3 = \text{constant.}$$

At a certain point of contraction, however, the electrons in a white dwarf would come to have so much energy on the average that it would pay them to combine with the protons to form neutrons, although a neutron has more mass/energy than a proton. This energy reversal explains why neutron stars form at the appropriate stage of a stellar collapse.

We now come to the point that is critical for the formation of black holes: *there is a limiting mass beyond which degeneracy pressure can no longer successfully balance gravity.* To see why this situation arises, we must look more carefully at the kinetic theory expression for the pressure of a gas. This pressure is exerted because momentum is transported from place to place by the constituent particles. One is thus concerned with the *flux* of momentum, and indeed one has the relation

$$P = npv$$

for the pressure P in terms of the concentration n of particles and their individual momentum p and velocity v. As the mass of the object increases and its radius decreases, the quantities n, p, and v will increase, and one can show that an equilibrium state is still possible. But this situation has a natural limit, since v will stop increasing once it approaches the velocity of light c. Because of this natural limit to the velocity with which the momentum can be transported, the degeneracy pressure can no longer increase sufficiently to balance the increased gravity, and the object must collapse. The existence of this limiting mass for a white dwarf or a neutron star was realized in the early 1930s by S. Chandrasekhar and L. Landau. For a neutron star its value depends somewhat on the still uncertain nuclear forces acting between the neutrons, but to within a factor better than 2 one has

$$M_c \sim 3M_\odot.$$

By great good fortune there are about 10^9 stars in our galaxy with a mass exceeding three solar masses, and by equal good fortune it is just the stars in this mass range which burn themselves out and then collapse in a time that is short compared to the age of the galaxy. We might expect, therefore, to find that a large number of black holes have formed in the galaxy, even when we allow for the fact that in many cases gravitational collapse may lead to a bounce followed by a supernova explosion in which some of the mass of the star is expelled. It seems unlikely that in every case the residue has a mass less than the Chandrasekhar limiting mass. Moreover, one has in the famous X-ray source Cygnus X-1 good (but not definitive) evidence that a black hole is actually present in the system.

Classical Black Holes

In the ideal case of a spherically symmetrical collapse, the system reaches the black hole state when its radius has shrunk to the critical value given by

$$R = \frac{2GM}{c^2}$$

$$\sim \frac{3M}{M_\odot} \text{ kilometers.}$$

Thus a solar mass black hole would have a radius of 3 kilometers. This critical or Schwarzschild radius gives the location of the event horizon of the black hole, so called because any event occurring beyond the horizon is invisible to an outside observer. The gravitational field at the horizon is so strong that light emitted at or within it cannot escape outwards. It will be observed that a neutron star, with its radius of about 10 kilometers, is itself not far from the black hole condition.

It is sometimes argued that the concept of a black hole is not self-consistent, on the grounds that, if light cannot reach an outside observer, the gravitational field of the material inside the horizon should also not be detectable outside. It is in fact true that gravitational waves emitted inside the horizon cannot escape, but the "D.C." gravitational field is not propagated as a wave and so is not filtered out by the horizon. Equally, if the material inside the horizon possessed a net charge, its Coulomb field would penetrate through the horizon, as indeed we would expect from an application of Gauss's theorem to a large sphere containing the black hole.

We thus arrive at the picture of a black hole which was accepted by theorists until 1974. We call it "classical" because the radical change in our thinking which was made in that year by Stephen Hawking depended on the introduction of quantum mechanical effects. In the classical picture, a black hole is the most passive type of object one can conceive of, capturing light or material particles by virtue of its gravitational field, but emitting nothing. This passivity could lead one to regard it as the end point of the evolution of a massive star. Hawking's great paper of 1974 completely destroyed this picture at one blow, for it showed that, far from being the most passive object known to man, a black hole is actually the most active. Because of quantum mechanical effects, a black hole radiates like a hot body, but with a temperature that increases as its mass decreases. The body thus radiates at an ever faster rate, ending its days in a violent explosion. My aim is to provide as simple a physical description of this great discovery as I can.

The Temperature of a Black Hole

The essential physical idea underlying Hawking's discovery is that the black hole distorts the quantum fluctuations of the vacuum and that this distortion has a thermal character. The existence of quantum fluctuations in the vacuum is a direct consequence of the Heisenberg Uncertainty Principle. This principle would be violated if the vacuum state were completely inactive, since all the dynamical variables of the system would then be zero and so known precisely. Instead, one has a zero-point energy associated with the quantum activity of the vacuum. This energy can be pictured in two somewhat different ways, each of which is helpful for understanding Hawking's result. In the first picture, one supposes that pairs of particles of the quantum field under consideration appear spontaneously and then reannihilate, leaving behind no residue of energy. In fact the energy budget resulting from this activity is compatible with the uncertainty principle if the pairs disappear on a time scale t related to their energy E by

$$t \sim h/E.$$

In the second picture, one decomposes the randomly fluctuating quantum field into modes of frequency v. Each mode then has a zero-point energy $\frac{1}{2} hv$. One also knows that in free space the number of modes per unit volume and per unit frequency range is proportional to v^2. One thus has an energy spectrum in the vacuum which depends on frequency like v^3.

We now consider how these vacuum fluctuations are distorted when they take place near the event horizon of a black hole. In the particle picture, we have to take into account the change in the energy of a particle resulting from its gravitational binding to the black hole. Classically the binding energy of a particle is always less than its kinetic plus rest mass energy, but in the extreme limit of a particle held permanently at rest close to the horizon, these two forms of energy become equal. This situation corresponds to the fact that even if all the atoms but one in the particle were converted into directed radiation, they could not blow the one remaining atom away from the horizon. This is just the condition that one has in a black hole. In these circumstances the total energy of the particle—rest mass plus binding—is precisely zero.

Once we understand this fact, it is not too difficult to go one step farther, to consider the total energy of a particle moving inside the horizon. In this case there are orbits for which the total energy is *negative,* because in those cases the binding energy dominates. This negative energy does not lead to paradoxes because it is a nonlocal quantity. A nearby observer would also be bound, and thus would not include the binding energy in his energy budget for the particle, unless he compared it with the energy at infinity which he took to be zero. This comparison would clearly be a nonlocal process. More-over, the total energy of a particle is classically a constant of the motion. Accordingly, if a classical particle falls freely into a black hole it cannot get onto a negative energy orbit, since it must be on a positive energy orbit when it is outside the black hole. Correspon-dingly, the mass of the black hole increases when it absorbs a classi-cal particle.

The situation is different in the quantum theory, as Hawking point-ed out. If we consider one particle of a pair that appears just outside the horizon in a quantum fluctuation, *this particle can be captured by the black hole into a negative energy orbit* by the quantum mechan-ical tunneling process (originally invoked to explain how a low ener-gy α-particle can escape from the strong binding of a radioactive nucleus). When this capture occurs the black hole *loses* mass. This mass/energy cannot disappear, of course, and must be given to the other particle of the original pair. Sometimes this particle also enters the black hole, in which case the final mass of the black hole is the same as its initial mass, and the process is not very interesting. Sometimes, however, the second particle moves away from the black hole, and *it now has enough energy to reach infinity,* where it can be observed. Thus, a distant observer will detect a flux of particles coming from the vicinity of the black hole, and these particles will be carrying away mass/energy from the black hole.

Despite the conceptual interest of this process of black hole evap-oration, one might at first think that the energy spectrum of the particles as observed at infinity would be complicated and essentially uninteresting. The really spectacular feature of Hawking's discovery is that this is not so. The black hole radiates as though it were a hot body at a temperature T given by

$$T = \frac{h}{16\pi^2 kGc^2} \left(\frac{1}{M}\right),$$

where k is Boltzmann's Constant. This radiation consists of all forms of matter and fields, since all of them are subject to the quantum fluctuations that drive the process and all of them are coupled to the gravitational field of the black hole. If we substitute the numerical values for the natural constants which appear in this formula we find

$$T \sim 10^{-7} \left(\frac{M_\odot}{M}\right) K.$$

Thus a black hole has a temperature, as Bekenstein had foreshadowed on other grounds. For a solar mass black hole the temperature is low, about 10^{-7} K, and as we shall see, the radiation rate is accordingly very slow. But if *low* mass black holes exist, perhaps formed as Hawking has suggested, in the early dense stages of the universe, their temperature would be *high*. For example, a 10^{15} gram black hole would have a temperature of 10^{11} K.

Once one has defined a temperature for a black hole it is natural to try and define an entropy also. One can in fact do this simply using the thermodynamic relation

$$dQ = TdS.$$

One then finds that the entropy is proportional to the *area* of the event horizon, as was again foreshadowed by Bekenstein. With these definitions for the temperature and entropy of a black hole, *Einstein's field equations imply that dynamical processes involving black holes obey all the laws of thermodynamics* (although the third law does depend on the validity of a plausible but unproved conjecture—Penrose's cosmic censorship hypothesis, which asserts inter alia that one cannot destroy the event horizon of a black hole). To develop a feeling for these remarkable results it is helpful to consider the thermal aspects that arise in a related situation in special relativity which also involves an event horizon, namely when an observer moves with uniform acceleration.

Uniformly Accelerated Observers

Motion with uniform acceleration occurs in special relativity when a constant force acts on a body moving in the direction z of the force. The velocity of the body asymptotically approaches the velocity of light relative to an inertial observer. Its world-line is a rectangular hyperbola in the z, t plane, and this hyperbolic motion was originally studied by Max Born in 1909 in his first published paper. It is a motion of interest for a number of reasons, and it has attracted a large literature, including a paper by Bondi and Gold which is concerned with the question of whether a uniformly accelerated charge radiates electromagnetic waves.

Our present interest in this motion stems from the fact that as the accelerated object moves away from a source of light at rest (in an inertial frame) on the z axis, there is a last moment at which light from the source would reach the accelerated object. Any light emitted after that moment would be unable to catch up with the object. Thus there is an event horizon present relative to the accelerating object. We therefore have a model of a black hole, but in the simpler context of special, rather than general, relativity.

Let us now consider the zero-point fluctuations of the vacuum as seen by a moving observer. This time it is more helpful to consider our second picture of these fluctuations, in which they are regarded as a random radiation field with a ν^3 spectrum. *This spectrum is Lorentz invariant.* In other words if one inertial observer is bathed in an isotropic ν^3 spectrum, then another inertial observer moving relative to the first with arbitrary velocity would also be bathed in an isotropic ν^3 spectrum. For this spectrum and this spectrum alone, all the Doppler shift and aberration factors cancel out when one makes a Lorentz transformation. This result, which is also important in cosmic ray physics, here guarantees that the quantum vacuum state does not pick out a preferred inertial observer who may be regarded as at rest. The quantum vacuum is Lorentz invariant.

Now we know from elementary dynamics that noninertial motion takes us outside Lorentz invariance. We can, for example, detect centrifugal forces in a rotating frame. We might therefore expect our accelerated observer to be able to detect his acceleration from a *change* in the spectrum of zero-point fluctuations relative to him. Now comes a great result, discovered by W. Unruh in 1976 (and foreshadowed by P. Davies in 1975). When the acceleration a is precisely uniform there is an additional spectrum that is precisely that of a heat bath at a temperature T given by

$$T = \frac{h}{4\pi^2 ck} \, a.$$

This is the same formula as for the Hawking temperature, if we replace the surface gravity GM/R^2 evaluated at the horizon of the black hole ($R = 2GM/c^2$) by the acceleration a. It is, in my opinion, one of the great formulas of physics, uniting as it does relativity (c), quantum theory (h), and thermodynamics (T). Yet it was discovered only four years ago!

The effect involved is small for all but the largest accelerations that arise in practice. For example, a temperature of 3 K would correspond to an acceleration of $10^{17}g$. One case where it may be important is the collision of two relativistic protons in a clashing beam accelerator. The protons would be brought to rest in a distance of about one fermi (10^{-13} centimeters), and in the uniform acceleration approximation each proton would find itself in a heat bath of about 10^{12} K.

As with the Hawking effect, it is not too surprising that noninertial motion is associated with some excess radiation. But why should this excess have a thermal spectrum when the motion is one of uniform acceleration? One clue to the explanation lies in a special property of this motion which is not quite obvious. As seen by an observer with this motion, the world is *time-independent*, a result that would not be true for generally accelerated motion, when, for example, inertial forces would be time-dependent. However, time-independence is necessary but not sufficient for thermal equilibrium. For example, one could have time-independent but nonequilibrium radiation in a cavity with smooth walls. The physicist knows what to do to achieve this equilibrium. He introduces a speck of dust into the cavity, or he roughens its walls. In other words, he introduces a coupling between the modes of the radiation field, relying on the fact that the only time-independent field that is stable to the switching on of such a coupling is one in thermal equilibrium. In our present case it turns out that the zero-point fluctuations of the vacuum are already in an equilibrated form, without the need for a speck of dust. This fact is related to the uniqueness of the vacuum for noninteracting fields. It thus comes about that the uniformly accelerated observer sees a heat bath at a definite temperature.

A Plausibility Argument for
Hawking Radiation

In order to connect this result with the thermal properties of black holes we now invoke the principle of equivalence, which was the centerpiece of Hermann Bondi's paper. Bondi used this principle to show that an observer in free fall sees no gravitational field in his locality. We want to use the principle in reverse. An observer not in free fall *does* see a gravitational field in his locality, even in special relativity. It is true that this field has a special character, namely that its gradient vanishes, as discussed by Bondi. Nevertheless, many physical processes occurring in a gravitational field are not very sensitive to this gradient. One can deduce the behavior of such processes in a gravitational field from the calculable effect on them of acceleration.

In this spirit we can argue that in the presence of a uniform gravitational field stretching so far that it contains an event horizon (corresponding to the building up of a potential difference $\Delta\phi = c^2$), the quantum fluctuations of the vacuum contain an additional thermal spectrum with temperature proportional to the gravitational field. Of course, the gravitational field of a black hole is *nonuniform*, since it decreases with increasing distance from the event horizon. Let us try to guess what effect this nonuniformity would have on the zero-point fluctuations. Consider an observer held at rest at some position outside the horizon of a black hole. Such an observer would not be freely falling. Rather, he would be accelerating relative to a local freely falling frame. Moreover, this acceleration would not change with time. We might then expect him to be in a heat bath with temperature proportional to the gravitational field at his location. We can repeat this argument at different locations to obtain a series of temperatures *that would be decreasing outwards* because of the nonuniformity in the gravitational field. Because of this temperature gradient there would be an outward heat-flow, and this is precisely the Hawking radiation from the black hole!

It might be objected that this argument would lead to a heat-flow from any nonuniform static gravitational field, such as that of a neutron star, or for that matter, of the earth. However, this is not so.

The event horizon plays a crucial role in the argument. To see this, consider for simplicity a body like the earth but with a perfect mirror coated on to its surface (the real earth would behave similarly in this problem). The in-falling zero-point modes of the vacuum would then be perfectly reflected to form a set of standing waves. The resulting field distribution would have a configuration reminiscent of a static atmosphere around a gravitating body. This configuration is called the Boulware state in the literature. It certainly involves no outgoing radiation, and so is in accordance with our normal physical intuition.

By contrast, we cannot support the zero-point modes against gravity right down to the event horizon of a black hole. If we tried to construct a static mirror at the horizon we would find that it would require an infinite force to prevent the mirror from being sucked into the black hole. Equally, we cannot prevent the zero-point modes from being sucked in, so no static atmosphere can now develop. It is *this* in-falling state that contains the outgoing Hawking radiation. The reason for the puzzling reversal of sign in the energy flow is the presence of the negative energy states inside the horizon which we discussed earlier.

This in-falling state is now called the Unruh state. It may be helpful to mention that there is yet another fundamental state for the radiation field. This is the Hartle-Hawking state. It describes the situation that would arise if the black hole were inside a large box with perfectly reflecting walls. In a steady state there would be built up a thermal radiation field in the box at the same temperature as the black hole, which would both emit and absorb radiation at the same rate. In this case the quantum field would be in the Hartle-Hawking state. The configuration is of great interest because if the box were large this equilibrium state would be *unstable* (a consequence of the temperature of the black hole being *inversely* proportional to its mass). However, for boxes smaller than a critical volume depending on the total energy available, the configuration would be stable.

Black Hole Explosions

We are now ready to consider the topic that gives this paper its title. Let us estimate the time scale for a black hole to radiate an appreciable fraction of its mass/energy in Hawking radiation. By Stefan's law the radiation rate per unit area integrated over all frequencies is proportional to the fourth power of the temperature. Since the area of the horizon is proportional to the square of the Schwarzschild radius, and the energy we are requiring to be radiated is proportional to the mass M of the black hole, we have finally for the time scale t_H

$$t_H \propto \frac{M}{T^4 R^2}$$

$$\propto M^3.$$

Thus the time scale depends sensitively on the mass. Moreover, as the radiation process continues, the mass of the black hole decreases, and the time scale for further radiation drops rapidly. In the final stages the time scale is so short that we are entitled to speak of an explosion.

To make this discussion more quantitative, we insert all the constants in the formula for t_H and find *for one radiated species*

$$t_H \sim 10^{10} \left(\frac{M}{10^{15} \text{ grams}} \right)^3 \text{ years.}$$

Thus for a solar-mass black hole the time scale would be ridiculously long, namely 10^{54} times the age of the universe, far longer even than the recently proposed lifetime of a proton, which is 10^{23} times the age of the universe. The Hawking radiation is clearly of no practical consequence for a solar-mass black hole. In fact such a hole would presumably accrete material faster than it radiates and so increase in mass.

The Hawking time scale would be of consequence, however, if there were in the universe mini-black holes of mass $\sim 10^{15}$ grams, since the time scale would then be of the same order as the age of the universe. (One would have to allow for the number of different species that would be radiated, but we are not concerned here with precise estimates.) Hawking has suggested that such objects might have been formed by a localized implosion in the early dense stages of the expanding universe. Such mini-black holes would be of great interest. They would, for example, have a Schwarzschild radius of 10^{-13} centimeters, just the radius of an elementary particle. The fact that the Hawking time scale for such an object is the age of the universe is itself a well-known numerical coincidence between cosmological and elementary particle quantities in a new guise.

Of course a radiation process that takes 10^{10} years is not exactly an explosion. Consider the situation when the mass of the black hole has decreased to 10^9 grams, however. The time scale is now one-tenth of a second if we still consider only one species. It seems fair to call this situation an explosion. The energy radiated in this time is about 10^{30} ergs, and if we think in terms of electromagnetic radiation, this energy would be mainly emitted in the form of 10^{13} electron volt γ-rays. Such γ-ray bursts are being searched for but none has been found as yet, a circumstance that places a limit on the number of mini-black holes formed in the early universe. Unsuccessful searches have also been made for radio and optical pulses that would be emitted, as Martin Rees has pointed out, by the electron-positron component of the Hawking radiation suddenly expanding in the magnetic field of the galaxy.

It would be very exciting if such pulses could be detected, so that one could explore observationally this remarkable phenomenon. Its most remarkable feature is the violence of the explosion. Even in the conservative case where only a single radiated species is considered, one is faced with an explosive power of 1 percent of the solar luminosity, but coming from an object of radius 10^{-19} centimeters or 10^{-90} of the volume of the sun! Such explosions justify my claim at the outset that what we thought to be the most passive objects in the universe are actually the most active.

REFERENCES

A more detailed account of the Hawking effect and of the thermodynamics of black holes, together with references to the original literature, may be found in the following papers.

BEKENSTEIN, J. D. 1980. *Physics Today 24.*

DAVIES, P. C. W., 1978. *Reports on Progress in Physics 41,* 1313.

Hawking, S. W., 1977. *Scientific American 236*, 34.

Sciama, D. W., 1976. *Vistas in Astronomy 19*, 385.

Sciama, D. W., P. Candelas, and D. Deutsch. 1981 *Advances in Physics, 30*, 327.

The articles of Bekenstein and Hawking are the most elementary and contain some historical remarks of great interest, coming as they do from the two chief founders of black hole thermodynamics.

6

THE BINARY PULSARS

Joseph H. Taylor

It is well known that about 330 pulsars have now been discovered. The number has doubled in the past three or four years, thanks to highly systematic surveys of the entire sky carried out at Molonglo (Manchester et al., 1978) and Green Bank (Damashek et al., 1978, and Damashek et al., 1982). I believe it is very unlikely that the number will change substantially in the next five years or so, because it is difficult to see how the survey sensitivities can be markedly improved without using prohibitive amounts of telescope time. Thus, we now have a relatively stable set of pulsar data for analysis and contemplation.

High quality timing measurements have been made for nearly three hundred of the pulsars, and have shown that most pulsars are exceedingly good timekeepers. All pulsars are found to be gradually slowing down—just as Tommy Gold predicted in 1968 (Gold, 1968)—but, after accounting for this easily measured, well-behaved effect, one can generally predict pulse arrival times to within a few ms or better, even after several years. Even the makers of digital quartz watches do not make such extravagant claims of precision for their products.

There are three exceptions to this simple picture of pulsar timing behavior, in each case because the pulsar in question is gravitationally bound into a binary star system. The pulse arrival times are therefore periodically Doppler shifted by the orbital motion, and the system may be studied by techniques similar to those used for spectroscopic binary stars. The binary pulsars are important because they provide the only direct method of determining neutron star masses. At the same time, they yield information on the evolution of close, high-mass binary stars, and, in one case, even probe the subtleties of relativistic gravitation theory.

The first known binary pulsar, PSR 1913+16, has been studied for over six years (Hulse and Taylor, 1975; Taylor et al., 1976; McCulloch et al., 1979; and Taylor et al., 1979), and I will bring you up to date on those observations momentarily. The second binary pulsar, PSR 0820+02, was recognized to be in an orbiting system more recently, in late 1978 (Manchester et al., 1980). As shown in Figure 6.1, its period had been uncharacteristically *decreasing* through 1977–1978; in late 1978 the period "bottomed out" and began in-

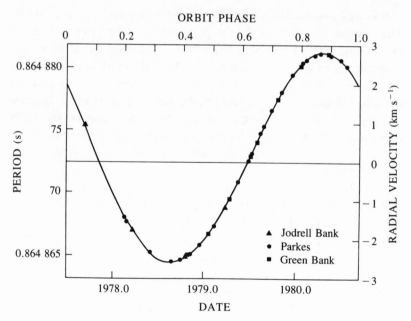

Figure 6.1. Period measurements of PSR 0820+02 during 1977–1980, and the corresponding radial velocities.

creasing once again, and it became evident that this pulsar was moving in a rather loosely bound binary system, with an orbital period of more than three years. Almost a full orbit has now been observed, and a least-squares fit to the velocity curve shown in Figure 6.1 yields (Taylor, 1981) an orbital period of $1154 \pm 30\ d$, an eccentricity close to zero, and an orbital radius (projected onto the line of sight) of $65 \pm 2\ R_\odot$. The pulsar mass function is $f(m_p) = (m_c \sin i)^3 (m_p + m_c)^{-2} = 0.0028\ M_\odot$. Although this expression does not yield the mass of either the pulsar (m_p) or the companion (m_c) uniquely, it does provide a useful constraining relation, which is plotted in Figure 6.2 as a function of $\cos i$ (i is the unknown angle between the plane of the orbit and the plane of the sky). The mass function is consistent with a pulsar of a solar mass or so, if the companion star has a somewhat smaller mass. There is not yet enough information available to allow a solution for either the inclination angle or the mass ratio.

The pulsar $0655+64$ was first observed (Damashek et al., 1978) in

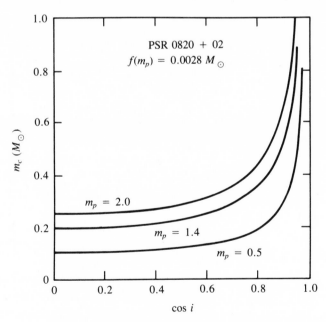

Figure 6.2. Mass of the companion of PSR 0820+02 as a function of $\cos i$, for each of three assumed pulsar masses.

early 1978. Subsequent observations in 1978–1979 showed its period to be variable, but the periodic trend of the variations was difficult to recognize—the reason being, as it turned out, that the orbital period $(24^h41^m\ 17.27^s\ \pm\ 0.08^s)$ is so close to one day (Figure 6.3). For this pulsar, too, the orbital eccentricity is very close to zero. The projected orbital radius is only 1.774 $R_\odot$, so if the companion is a main sequence star, the pair are very close and tidal effects will be very important in the evolution of the orbit. Blandford and DeCampli (1981) have argued that the companion is more likely to be a white dwarf. In any case, the mass function, $f(m_p) = 0.0712$, is consistent with a companion mass in the range ~0.5 to ~1.5 $M_\odot$, as shown in Figure 6.4.

It is interesting to note that both PSR 0820+02 and PSR 0655+64 are well out of the galactic plane ($b = +25°$ and $+21°$, respectively) and have rather small dispersion measures indicating distances of ~1 kpc and ~0.3 kpc. Thus, unless the companions are neutron stars or black holes, optical studies may prove to be fruitful. Such work would be very important, but must await the accurate pulsar position

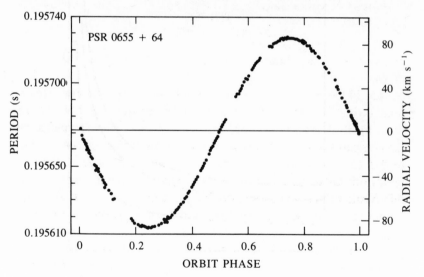

Figure 6.3. Period and radial velocity of PSR 0655+64 as a function of orbital phase.

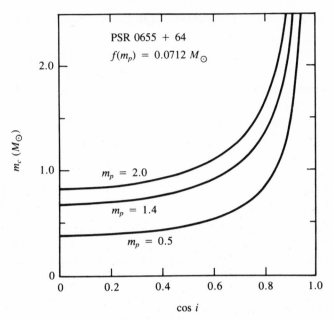

Figure 6.4. Mass of the companion of PSR 0655+64 as a function of cos i, for each of three assumed pulsar masses.

measurements that should be available from timing data within a year or so.

PSR 1913+16 has the shortest orbital period of the three, approximately $7^h\,45^m$, and a rather large orbital eccentricity, $e = 0.617$. The mass function, $f(m_p) = 0.1312\,M_\odot$, is consistent with the pulsar and the companion star both having masses in the range $\sim$0.5 to $\sim$2$M_\odot$. In this case, however, additional information is available: the maximum orbital velocity is $v/c \sim 10^{-3}$, and as a consequence several relativistic effects are of measurable magnitude (Taylor et al., 1976; Taylor et al., 1979; and Blandford and Teukolsky, 1976). By far the largest of these magnitudes, is the secular advance of the longitude of periastion of the orbit. This angle is observed to increase at the astonishing rate of 4.226 degrees per year, or some thirty-five thousand times faster than Mercury's 43 arc sec per century. If, as appears to be the case, the rotation of the orbit is entirely due to the general relativistic effect, then the rate establishes the total mass of the system at $m_p + m_c = 2.826 \pm 0.001 \, M_\odot$.

The next largest relativistic effect is the time dilation arising from second-order Doppler shift and gravitational red shift. Because of the large orbital eccentricity, these effects give rise to an observable, periodically varying delay of amplitude ~4 ms. Because this delay depends on a different combination of the two masses than does the rate of periastron advance, it is possible to solve the relevant expression (Blandford and Teukolsky, 1976) explicitly for the pulsar and companion star masses. Such a procedure, outlined by Saul Teukolsky in the next chapter, yields $m_p = 1.43 \pm 0.07 \, M_\odot$, $m_c = 1.40 \pm 0.07 \, M_\odot$.

General relativity requires that a pair of orbiting masses must loose

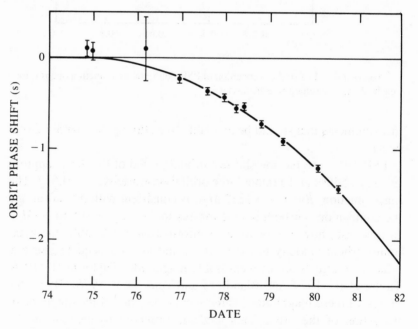

Figure 6.5. Observed accumulation of excess orbital phase for PSR 1913+16, relative to a hypothetical orbit of assumed constant period. Excess phase expected because of gravitational radiation is indicated by the curvature of the parabola drawn through the points. First-order slope and intercept of the parabola have been adjusted for best fit by the method of least squares.

energy by radiating gravitational waves. The observable consequence will be a secular decrease of orbital period as the two objects spiral closer together. For the PSR 1913+16 system, the expected rate of period decrease (Taylor et al., 1979; Peters and Matthews, 1963; and Wagoner, 1975) is $(-2.38 \pm 0.02) \times 10^{-12} s \ s^{-1}$. As shown in Figure 6.5, our observations are currently in very good agreement with this prediction. We take this agreement as compelling evidence for the existence of gravitational waves.

REFERENCES

BLANDFORD, R. D., AND S. A. TEUKOLSKY, 1976. *Astrophys. J. 205*, 580.

BLANDFORD, R. D., AND W. M. DECAMPLI, 1981. In *IAU Symposium No. 95* "Pulsars," W. Sieber and R. Wielebinski, eds., Reidel, Dordrecht, pp. 371–377.

DAMASHEK, M., J. H. TAYLOR, AND R. A. HULSE, 1978. *Astrophys. J. (Letters) 225*, L31.

DAMASHEK, M., P. R. BACKUS, J. H. TAYLOR, AND R. K. BURKHARDT, *Astrophys. J. (Letters)*, forthcoming.

GOLD, T., 1968. *Nature 218*, 731.

HULSE, R. A., AND J. H. TAYLOR, 1975. *Astrophys. J. (Letters) 195*, L51.

MANCHESTER, R. N., A. G. LYNE, J. H. TAYLOR, J. M. DURDIN, M. I. LARGE, AND A. G. LITTLE, 1978. *Mon. Not. Roy. Astron. Soc. 185*, 409.

MANCHESTER, R. N., L. M. NEWTON, D. J. COOKE, AND A. G. LYNE, 1980. *Astrophys. J. (Letters) 236*, L25.

MCCULLOCH, P. M., J. H. TAYLOR, AND J. M. WEISBERG, 1979. *Astrophys. J. (Letters) 227*, L133.

PETERS, P. C., AND J. MATTHEWS, 1963. *Phys. Rev. 131*, 435.

TAYLOR, J. H., 1981. In *IAU Symposium No. 95* "Pulsars," W. Sieber and R. Wielekinski, eds., Reidel, Dordrecht, pp. 361–369.

TAYLOR, J. H., R. A. HULSE, L. A. FOWLER, G. E. GULLAHORN, AND J. M. RANKIN, 1976. *Astrophys. J. (Letters) 236*, L25.

TAYLOR, J. H., L. A. FOWLER, AND P. M. MCCULLOCH, 1979. *Nature 227*, 437.

WAGONER, R. V., 1975. *Astrophys. J. (Letters) 196*, L63.

7

THE IMPORTANCE OF THE BINARY PULSAR FOR GENERAL RELATIVITY

Saul Teukolsky

It is a great pleasure to include in the present volume this short essay about the importance of the binary pulsar for general relativity, for it is well known that Tommy Gold's name is intimately associated with the pulsar story. As soon as word of the discovery of the binary pulsar by Hulse and Taylor (1975) reached Cornell, Tommy of course organized a group of people to discuss what the implications might be. In this paper, I will briefly summarize the importance of this discovery today, confining my remarks to general relativity, although there are also significant implications for pulsar theory, the theory of the evolution of binary systems, and so on.

Certainly the most exciting aspect of this system is the possibility of demonstrating the existence of gravitational waves, with properties in accordance with Einstein's theory. Two objects in orbit about each other have a time-varying quadrupole moment, and so radiate gravitational waves. The loss of energy leads to a decrease in the radius of the orbit and hence a decrease in the orbital period, $\dot{P}$.

Now it is possible to imagine other astrophysical effects that can change an orbital period, such as mass loss. Also, an acceleration of the center-of-mass of the system shows up in the arrival time of pulses as an apparent $\dot{P}$. How can we be sure that an observation of $\dot{P}$ must be due to gravitational radiation?

To answer this question, let me focus on four key parameters of the binary system which must first be measured: the masses of the pulsar and its companion, M_1 and M_2; the semi-major axis of the pulsar's orbit, a_1; and the sine of the inclination angle of the orbital plane to the line of sight, $\sin i$.

Two relations between these four parameters are readily determined. The first-order Doppler shift (simply due to the light travel time across the orbit) gives

$$x = a_1 \sin i. \tag{1}$$

The orbital period P is also readily measured. Define $\wp = P/2\pi$. Then Kepler's Third Law gives

$$\wp^2 = \frac{(a_1 + a_2)^3}{M_1 + M_2}. \tag{2}$$

Combining (1) and (2), we get

$$\frac{x^3}{\wp^2} = \frac{(M_2 \sin i)^3}{(M_1 + M_2)^2}, \tag{3}$$

called the mass function of the system.

A third relation follows from the general relativistic prediction for the periastron advance:

$$\dot{\omega} = \frac{3(M_1 + M_2)^{2/3}}{\wp^{5/3}(1 - e^2)}, \tag{4}$$

where e is the eccentricity.

Note that we are using units with $c = G = 1$, so that $M_\odot = 1.476625$ km. The mass of the sun in grams is not nearly as precisely known because of the uncertainty in G, and the accuracy of the observations has reached the point that this uncertainty makes a difference! For example, the measured value $\dot{\omega} = 4.2256$ deg/yr implies, with the known values of P and e,

$$M_1 + M_2 = 2.8274 \, M_\odot \tag{5}$$

a value that cannot be quoted as precisely in grams. Considering the system is about 5 kpc away, this is a remarkable indication of the precision of pulsar timing.

The combined effects of the second-order Doppler shift and the gravitational red shift show up as another parameter in the timing equation,

$$\gamma = \frac{\wp^{1/3} e \, M_2 \, (M_1 + 2M_2)}{(M_1 + M_2)^{4/3}} \tag{6}$$

Note that formally $\gamma/x \sim v/c \sim 10^{-3}$, where v is the orbital velocity. Thus γ appears at first sight to be readily measurable. However, if $\dot{\omega}$ were zero, γ would be exactly degenerate with x and would be unmeasurable (Blandford and Teukolsky, 1975). One has to wait for the periastron position to change before γ is measurable. This measurement has now been accomplished to an accuracy of several percent, as Joseph Taylor reports in this volume.

Assume for the moment that nongeneral relativistic sources of periastron advance are negligible (e.g., due to a quadrupole moment of the companion). In this case, one has four equations that can be solved for the four parameters M_1, M_2, a_1, and $\sin i$. Now general relativity predicts that

$$\dot{P} = - \frac{M_1 M_2}{\wp^{5/3} (M_1 + M_2)^{1/3}} \frac{192\pi}{5} \frac{(1 + 73 \, e^2/24 + 37 \, e^4/96)}{(1 - e^2)^{7/2}} \tag{7}$$

due to gravitational radiation emission. In order of magnitude, $\delta P \sim 10^{-4} s$ per yr.

Suppose one measures $\dot{P}$, and suppose it agrees with the prediction of (7). It would be a miracle if general relativity were not true, and yet other effects conspired to give exactly the value of (7). Moreover, it would be a miracle if other possible contributions for example to $\dot{\omega}$ also conspired to give exactly the correct value.

As Joseph Taylor mentioned, the measured value of $\dot{P}$ agrees with

(7) to within the measurement error of around 20 percent, and the error should decrease quite rapidly in the future.

One can go further. As the timing accuracy improves, it becomes possible to measure further relativistic effects (Blandford and Teukolsky, 1975; Epstein, 1977), such as the Shapiro time-delay of signals as they cross the orbit on the way to the earth, and various post-Newtonian periodic deviations from elliptic motion, never before verified in the solar system. In general relativity, each of these terms contains a known different combination of M_1, M_2, a_1, and $\sin i$. As more and more of these terms are measured (and it looks as if we are on the verge of being able to do so), and if they agree with the general relativistic predictions, then the case for general relativity will become increasingly watertight.

An interesting relativistic effect that occurs if the pulsar spin axis is not exactly perpendicular to the orbital plane is the geodetic precession. (The Lens-Thirring precession is smaller by about 10^{-3} here). The precession angular velocity, Ω_G, is (Borner et al., 1975)

$$\frac{\Omega_G}{\dot{\omega}} = \frac{M_2(4M_1 + 3M_2)}{6(M_1 + M_2)^2} , \qquad (8)$$

or about $\dot{\omega}/3 \sim 1$ deg/yr. No evidence for this effect has shown up yet, for example as a change in pulse shape.

Most other theories of gravitation predict dipole gravitational radiation as well as quadrupole. This radiation would be stronger than the quadrupole emission by a factor $\sim c^2/v^2 \times$ structure factors, where $c^2/v^2 \sim 10^6$. This possibility is almost certainly ruled out (Will, 1977). Thus the binary pulsar has already had a significant impact on strengthening our belief that Einstein was right after all.

REFERENCES

BLANDFORD, R., AND S. A. TEUKOLSKY, 1975, *Astrophys. J. Lett. 198,* L27; see also, Brumberg, V. A., Ya. B. Zel'dovich, I. D. Novikov, and N. I. Shakura, 1975, *Astr. Letters 1,* 5.

BLANDFORD, R., AND S. A. TEUKOLSKY, 1976, *Astrophys. J. 205,* 580.

Borner, G., J. Ehlers, and E. Rudolph, 1975, *Astron. Astrophys. 44*, 417.

Epstein, R., 1977, *Astrophys. J. 216*, 92.

Hulse, R. A., and J. H. Taylor, 1975, *Astrophys. J. Lett. 195*, L51. The latest observations are summarized by Taylor in the previous paper in this book.

Will, C. M., 1977, *Astrophys. J. 214*, 826, and references therein.

8

ON SEEING A NEUTRON STAR

David J. Helfand

In the autumn of 1967, short, regular bursts of celestial radio emission were first recorded at Cambridge and, within a year, Tommy Gold had convinced most astronomers that a neutron star was responsible (Gold 1968). The observation of a pulsar in the Crab Nebula shortly thereafter confirmed his hypothesis. In the intervening dozen years, we have accumulated a wealth of data on pulsars over twenty decades of the electromagnetic spectrum and, as Professor Sciama notes in Chapter 5, over the spectrum of fields in theoretical physics as well. However, we have yet to *see* a neutron star in the sense that we see other stars—by detecting thermal radiation from the stellar surface (Greenstein and McClintock, 1974). Deprived of similar information on the sun and other stars, restricted, for example, to observing coronal X rays and nonthermal radio bursts, our progress in understanding the major constituent of our galaxy would clearly have been severely hampered. This analogy is not as tenuous as it at first may seem, for the radio emission from pulsars is as energetically insignificant as are solar coronal X rays; both represent $\sim 10^{-6}$ of the total energy being lost by the underlying star.

How, then, should one undertake a search for neutron star surface emission? The observational constraints are severe. If a typical neutron star had a surface temperature similar to that of the sun, it would have a visual magnitude of ~32 at a distance of 100 pc. If it were ten or twenty times hotter with a thermal spectrum that peaked in the ultraviolet, our search would be equally fruitless owing to the extremely high opacity of the interstellar medium in this wavelength range. We know, however, that most radio pulsars do not have temperatures $\gtrsim 10^7$ K from the fact that none of the 330 known pulsars is coincident with any of the 350 X-ray sources found in the 2—10 keV all-sky X-ray surveys conducted in the early 1970s. What remains, then, is the temperature range of a few hundred thousand to a few million degrees. With the launch of the Einstein Observatory soft X-ray telescope in 1978, we have increased our sensitivity to objects emitting in this 0.1—5 keV band by several orders of magnitude and are at last in a position where we can hope to see a neutron star.

There are two primary sources of thermal energy available to a neutron star: the reservoir of heat remaining from its formation at the center of the supernova explosion that created it, and the subsequent dissipation of its rotational kinetic energy as it gradually spins down. From an initial temperature of $\sim 10^{11}$ K, a young neutron star is expected to cool at a rate such that its surface temperature will remain above 10^6 K for from 10^3 to 10^4 years. Searches of the center of young supernova remnants, then, should reveal point sources of the soft X-ray emission characteristic of such temperatures. For the older radio pulsars, a variety of mechanisms have been proposed which would keep the surface above 10^5 K for 10^6—10^7 years (see Helfand et al., 1979, for a detailed review). Briefly, these mechanisms include: heating of the magnetic polar cap via bombardment of relativistic particles flowing inward from the pair creation zones in the magnetosphere, the frictional coupling between the slowing crust and more rapidly spinning superfluid interior, the release of strain and gravitational energy in a starquake, the magnetic dipole radiation of neutrons circulating in the superfluid vortices of the core (Huang et al., 1980), and, for the oldest stars, accretion from the interstellar medium. A survey of radio pulsars for thermal emission in the temperature range 2—10 × 10^5 K will thus provide important constraints on several aspects of pulsar emission models and theories of neutron star structure. We report below on the current status of observational programs designed to detect both young and old neutron stars.

A search has been conducted using the imaging instruments on board the Einstein Observatory for central point-source emission in the seven historical supernova remnants (CasA, Kepler, Tycho, 3C58 [SN1181], the Crab, SN1006, and RCW86 [SN185]) in addition to about forty other radio remnants within ~ 8 kpc. Prior to these observations, calculations spanning a variety of neutron star masses, equations of state, and magnetic field strengths yielded a lower bound to the predicted temperatures for a 10^4 year-old neutron star of $>2 \times 10^6$ K (Tsuruta, 1979); such objects would have been easily detectable in virtually all of the remnants studied. In fact, at most four central point sources have been discovered. Stimulated by the possibility of observational progress, several groups have reinvestigated the cooling problem during the last two years (Glen and Sutherland, 1979, 1980; Lamb and van Riper, 1979, 1980). Incorporation of certain general relativistic effects that had been neglected in earlier work, coupled with improved high energy, nuclear, and solid-state physics parameters has led to a substantial reduction in predicted temperatures, although agreement among the various groups has not yet been achieved (Tsuruta, 1980). Lower limits to predicted temperatures for $1 M_{\odot}$ magnetic ($B \sim 10^{12}$ G) stars assuming a stiff equation of state now lie in the range $1—2 \times 10^6$ K for a three hundred-year-old object and .5 to 1.5×10^6 K for a 10^4-year-old star. As has been recognized for some time, the existence of a pion condensate stellar core (or, perhaps, a quark-dominated interior) will lead to much more rapid cooling, allowing the star to reach temperatures of $<3 \times 10^5$ K in a few hundred years.

The only historical SNR known to contain a neutron star is the Crab Nebula. Its central pulsar has long been known to be a source of X-ray emission, but it was established through lunar occultation measurements in 1974 (Wolff et al., 1975; Toor and Seward, 1977) that most of the X rays are pulsed, with a double-peaked profile very similar to that seen in the radio, optical., and gamma-ray regimes, and are thus presumably of nonthermal origin. Einstein observations with 5" resolution have revealed that soft X-ray emission persists throughout the pulse period, reaching a minimum of about 1 percent of the pulse peak ~40° before the main pulse (Harnden et al., 1979). Interpreted as surface thermal emission, these X rays imply a temperature of ~2 × 10^6 K for this thousand-year-old star. In the absence of spectral information on this persistent emission and in light of the recent observation that polarized optical emission also persists throughout all pulse phases with a similar peak-to-minimum ratio (Smith, 1981), however, it seems most reasonable to treat this value as a temperature upper limit. The remnant 3C58 (SN1181) has a very similar radio morphology (centrally peaked emission, polarized filaments, etc.) to that of the Crab, and a 1' resolution X-ray image also shows a bright, centrally peaked distribution. The detection of a weak point source at the center of this remnant has recently been reported (Becker, 1980), although temporal and spectral information are not yet available. As for the other historical remnants, no central sources are seen with temperature upper limits varying from <2 × 10^6 K for Tycho and Kepler to <0.7 × 10^6 K for SN1006 (Helfand et al., 1979). All of these values are close to the lower bounds of the newly calculated cooling curves, although none of the predictions can comfortably accommodate a neutron star in SN1006 unless it underwent rapid early cooling as the result of a pion core.

Likewise, only two of the remaining forty remnants studied show evidence of a central source. The only other known pulsar-SNR association is Vela and, again, an X-ray source coincident with the pulsar is seen. A high resolution observation has shown that ≳50% of the flux from this source is in an extended nebula with a spectrum and morphology similar to that of the synchrotron nebula surrounding the Crab pulsar. The X rays that come from the point source centered on PSR 0833−45 itself are unpulsed (Harnden, 1980). Interpreted as thermal emission, they require a temperature of ~1 × 10^6 K although, again, a nonthermal origin cannot be excluded. In the young southern remnant RCW103, a point source is seen centered in a typical circular remnant shell; the implied temperature of a 10 km blackbody at the remnant distance of 2.6 kpc is ~2.5 × 10^6 K (Tuohy and Garmire, 1980). For the other three dozen remnants surveyed, however, no excess emission is apparent, leading to the conclusion that either the standard cooling calculations overestimate expected temperatures or many supernovae (SN) leave no neutron star remnants.

The latter possibility is supported by an independent line of reasoning employing the same data set (Helfand, 1981). For the two known radio pulsar-SNR associations, we see an X-ray synchrotron nebula surrounding the pulsar. This nebular emission is ultimately powered by the rotational energy loss, $\dot{E}$, of the central neutron star. We might therefore expect that the X-ray luminosity of such nebulae will roughly scale as $\dot{E}$ to some power n. The Vela nebula has a linear size $\sim 1/5$ and a luminosity $\sim 10^{-4}$ that of the Crab, while the corresponding pulsars have a ratio of $\dot{E}$ of $\sim 10^{-2}$. If there were a rapidly rotating, magnetized neutron star (i.e., a young radio pulsar) at the center of any of the historical remnants, we would expect to detect it easily. Using the two known examples to assign n a value of 2, our lack of detection of any such central synchrotron nebulae in the historical remnants allows us to place an upper limit on the value of $\dot{E}$ for their central neutron stars of $\sim 10^{-2}$ to $10^{-3} \dot{E}_{\text{Crab}}$. Such neutron stars cannot be typical young pulsars; they must have very low magnetic fields and/or relatively long periods (~ 0.3 s) and, as such, would lead to a $P\text{-}\dot{P}$ diagram for pulsars different from that which is observed. We thus conclude that a significant fraction of SN do not leave rapidly spinning neutron star remnants. Such an observation is consistent with current ideas regarding models of Type I SN (of which Tycho, Kepler, and SN1006 may all have been examples), but serves only to worsen the already significant discrepancy between the galactic SN rate and the pulsar birthrate.

The second part of our study concerning thermal emission from neutron stars is designed to delimit the heating (rather than cooling) mechanisms operative in radio pulsars. Eighteen objects have been observed to date and, in six instances, soft X-ray sources have been detected within the combined X-ray and radio positional uncertainties; less than 1 coincidence would be expected by chance at our flux limit of $\sim 10^{-13}$ erg cm^{-2}s^{-1}. The strongest coincident source is located within $3'' \pm 5''$ of PSR 1055-52. No optical object is seen to $m_v \approx 20$ within the X-ray error circle, implying an X ray-to-optical luminosity ratio for the object of >400:1. The only known X-ray sources with such an extreme ratio are the X-ray binary systems;

however, to fit this source within the luminosity range of 10^{36}-10^{38} erg s^{-1} characteristic of that population would require a distance of 10 − 100 kpc (at $b \sim 10°$), placing the source far outside of the galactic disk. In addition, the soft spectrum and lack of appreciable low energy absorption argue strongly against such an interpretation. The X-ray flux and spectrum are consistent, however, with what is expected from a 16 km blackbody radiating at a temperature of $\sim 1 \times 10^6$ K at the pulsar distance of 1100 pc. When final instrument calibrations are complete, we may be able to fit the observed X-ray energy distribution to a Planck function and, through comparison with the flux, measure the neutron star radius. Deviations in the spectrum from a simple blackbody law will provide information on the structural properties of the surface material and/or the temperature gradients over the surface of the star. The rather high ratio of the pulsar's spindown energy which is being converted to thermal energy ($\sim 1\%$) will place important constraints on the various heating mechanisms noted above. The observation that the emission is not modulated at the pulsar rotation frequency (to the $\sim 15\%$ level) may also provide a limit to the inclination angle of the magnetic poles to the rotation axis. Assuming that the poles will be at a different temperature than the magnetic equator as a result either of polar cap heating or a nonuniform thermal conductivity induced by the magnetic field, we can place a limit on the angle of $\lesssim 20°$. This is an important measurement for this particular pulsar, in that its radio pulse profile is very similar to that of the Crab (McCulloch et al., 1976), with a strong interpulse appearing nearly halfway through the period. If the claim of a small inclination angle can be substantiated, it will lend strong support to the one-pole model of the interpulse emission.

Most of the other X-ray sources coincident with pulsars have spectra and fluxes consistent with considerably lower temperatures than PSR 1055−52; the emission from PSR 1642−03 implies a temperature of only 2.9×10^5 K, although the ratio of thermal to spindown energies is also a few percent. Other possible detections lie between 3 and 8×10^5 with upper limits ranging down to a low of $<2 \times 10^5$ K for PSR 1929+10; thermal to spindown energy ratios range from $\sim 100\%$ for PSR 0031−07 (casting serious doubt that the identification with the pulsar is correct), to $<0.1\%$ for PSR 0655+54, the third binary pulsar. These data provide several interesting constraints on neutron star and pulse emission models. The generally low temperatures require, in the context of frictional heating scenarios (Harding et al., 1978), that the pulsars have low masses ($\sim$0.5 $M_\odot$). This is in conflict with the measured masses of neutron stars in binary systems which cluster around 1.4 $M_\odot$. The strong temperature upper limit on the 24^h binary pulsar will, when its mass has been established, provide a useful confrontation between theory and observation: it may show, for example, that the assumption, implicit in the frictional heating calculations, of no significant superfluid vortex line pinning must be modified. For polar cap emission models, the limit on observed cap temperatures yields a value for the parameter representing the binding energy of the nuclei in the crust and speaks to issues that include the degree of positron trapping in the outer magnetosphere and the importance of synchrotron losses undergone by particles headed back toward the surface. The potential gain for the study of neutron star equations of state accruing from the radius measurement for PSR 1055−52 described above is manifest.

The detection of unpulsed soft X-ray emission from some old pulsars adds yet another piece of evidence to the case for the lack of young pulsars in most of the SNR. We may be forced by the growing SN rate/pulsar birthrate discrepancy to find alternative sites for neutron star formation. Having established the orthodoxy of SN as pulsar birthplaces over a decade ago, Tommy Gold characteristically forged ahead to later suggest ''fizzlers'' as a quieter way of making neutron stars. It may be that, once again, the observations are now beginning to catch up with one of his apocryphal ideas.

REFERENCES

BECKER, R., private communication.
GLEN, G., AND P. G. SUTHERLAND, 1979. *BAAS 11*, 779.
GLEN, G. AND P. G. SUTHERLAND, 1980. Submitted to *Ap. J.*
GOLD, T., 1968. *Nature 218*, 731.
GREENSTEIN, G. AND J. E. MCCLINTOCK, 1974. *Science 185*, 487.
HARDING, D., R. A. GUYER, AND G. GREENSTEIN, 1978. *Ap. J. 222*, 991.

Harnden, F. R.. May 1980 talk presented at workshop on "Stellar Collapse, Supernovae, and Neutron Star Formation," Theoretical Physics Institute Santa Barbara, California.

Harnden, F. R., B. Buehler, R. Giacconi, J. Grindlay, P. Hertz, E. Schreier, F. Seward, H. Tananbaum, and L. van Speybroeck, 1979. *BAAS 11*, 789.

Helfand, D. J. 1981. *In IAU Symposium No. 95*, "Pulsars," W. Sieber and R. Wielebinski eds., Reidel, Dordrecht, pp. 343–350.

Helfand, D. J., G. Chanan, and R. Novick, 1979. *Nature, 283*, 337.

Huang, J. H., R. E. Lingenfelter, Q. H. Peng, and K. L. Huang, 1980. Submitted to *Nature*.

Lamb, D. Q., and K. A. van Riper, 1979. *BAAS 11*, 779.

Lamb, D. Q., and K. A. van Riper, 1980. Preprint.

McCullock, P. M., P. A. Hamilton, J. G. Ables, and M. M. Komesaroff, 1976. *MNRAS 175*, 718.

Smith, F. G. 1981. *In IAU Symposium No. 95*, "Pulsars," W. Sieber and R. Wielebinski eds., Reidel, Dordrecht, pp. 221–234.

Toor, A., and F. D. Seward, 1977. *Ap. J. 216*, 560.

Tsuruta, S., 1979. *Phys. Reports 56*, 237.

Tsuruta, S., 1981. *Proc. IAU Symposium No. 95*, "Pulsars," W. Sieber and R. Bonn, Wielebinski, eds., Reidel, Dordrecht, pp. 331–337.

Tuohy, I. and G. Garmire, 1980. *Ap. J. (Letters) 239*, L107.

Wolff, R. S., H. L. Kestenbaum, W. Ku, and R. Novick, 1975. *Ap. J. (Letters) 202*, L77.

PART III

THE SOLAR SYSTEM

Plate 3. The rings of Saturn. Courtesy, JPL/NASA.

9

RESONANCES AND RINGS IN THE SOLAR SYSTEM

Peter Goldreich

It is a great pleasure to dedicate this paper to Tommy Gold on the occasion of his sixtieth birthday. I first met Tommy twenty-one years ago during my senior year at Cornell (I majored in engineering physics), when I was looking around for a summer job. He not only hired me but subsequently encouraged me to stay on at Cornell as a graduate student in physics. During my graduate student days I lived in an attic apartment in Tommy's house. These three years were exciting for me. My frequent interactions with Tommy shaped both the content and style of my work. Many, if not most, of the problems I have investigated are ones on which Tommy has also left his mark. With so much overlap in scientific interests, it is not surprising that Tommy and I have sometimes been in competition. What is perhaps not well known is that when I lived in his house we competed athletically as well as intellectually. Our physical competitions offered me one opportunity, probably unique in graduate student/thesis adviser relations, which I eagerly seized. In 1963, I was a medal winner at the NCAA judo championships. Shortly thereafter, Tommy expressed doubt about the efficiency of judo. We were standing in his kitchen at the time and I strangled him on the spot.

Initially, I had intended to survey the great variety of dynamical resonances in the solar system, but upon reflection, I decided that this subject is too broad to treat properly in a single short paper. Therefore, I shall concentrate on planetary rings and the role that satellite resonances play in determining their structure. Before I settle down to the main body of my discussion, I shall offer a few remarks of a more general nature concerning solar system resonances and Tommy's role in elucidating their nature.

Roy and Ovenden (1954, 1955) concluded that there is a preference for low order commensurabilities among the orbital periods of the satellites of Jupiter and Saturn which is inconsistent with the assumption of a random distribution of orbital periods for these bodies. Tommy introduced me to this problem and suggested that its resolution might involve the tides. I proved that the observed commensurabilities would not be disrupted by tidal torques (Goldreich, 1965). The direct gravitational interactions between pairs of satellites redistribute the angular momentum that the tidal torque adds to each satellite's orbit in just the correct proportions to preserve the commensurabilities. The obvious hypothesis is that the satellite orbit commensurabilities arose as a consequence of tidal evolution. A spectacular extension of these ideas to the tidal heating of Io is due to Stan Peale, another of Tommy's former students (cf. Peale et al., 1979).

Radar studies of Mercury using Cornell's Arecibo radio telescope led to the unexpected discovery that the planet's spin was not synchronous with its orbital revolution (Pettengill and Dyce, 1965). It was immediately recognized by Peale and Gold (1965) that the solar tidal torque and Mercury's eccentric orbit must together be responsible for the nonsynchronous spin. The key idea was provided by Colombo (1965). He pointed out that there were other spin resonances besides the synchronous one and correctly surmised that Mercury's spin period is two-thirds as long as its orbital period. The explanation of how Mercury's spin was trapped in this resonant state was provided by Goldreich and Peale (1966).

The Kirkwood gaps in the number density versus semi-major axis distribution of the asteroids are another and unexplained example of solar system resonances. These gaps occur where the asteroidal orbit periods are commensurate with Jupiter's orbit period. Dissipative effects in the asteroid belt are, at least at present, too weak to contribute to gap formation. I favor the view that the Kirkwood gaps may be understood as a result of the gravitational N-body problem. Possibly the orbital eccentricities of asteroids originally located in the gaps grew until these orbits crossed or closely approached those of Mars or Jupiter and a catastrophic interaction occurred. Of course, this is mere speculation but perhaps it is worth pursuing so that I can indicate a potential mechanism for the eccentricity increase. Near each gap there are a number of separate resonances with slightly different frequencies. The crucial point is that the widths of the individual resonances are greater than the frequency separations so that there is resonance overlap. We know from studies in other fields that overlapping resonances lead to chaotic orbital evolution (Chirikov, 1979). It is worth noting that the resonances in satellite systems are relatively narrower and more widely spaced than those in the asteroid belt, and that overlap does not occur.

Observational Properties of Planetary Rings

Nature has presented us with three systems of planetary rings. They are not at all similar. Jupiter possesses a faint ring of very low optical depth. Saturn has bright, broad rings separated by narrow gaps. The rings of Uranus are dark, narrow, and widely spaced.

All three ring systems are located close to the parent planet whose tidal gravity prevents the ring material from forming into satellites.

Saturn's Rings

Observations at visual wavelengths reveal two bright rings, A and B, separated by a dark gap, the Cassini division. A third faint ring, C, lies inside the bright ones. The mean normal optical depths of the rings are $\tau_A \sim 0.3$, $\tau_B \sim 0.7$, and $\tau_C \sim 0.1$ (Cook and Franklin, 1958). An upper limit to the ring thickness of 3 km is deduced from the variation of ring brightness during the earth's periodic passage through the ring plane (Bobrov, 1970).

Since near infrared reflectance spectra show absorption features due to water frost (Pilcher et al., 1970), the ring particles are at least coated with ice.

Measurements of the thermal emission at far infrared wavelengths yield brightness temperatures, T_B, consistent with the equilibrium temperature, T_E, for highly reflecting particles at Saturn's distance from the sun (Murphy, 1974). T_B ranges from ~70 K near the time the sun crosses the ring plane to ~90 K at maximum opening angle of 26°.

The rings are very weak emitters at radio wavelengths. Interferometric maps yield $T_B < 10$ K in the centimeter wavelengths range (Berge and Muhleman, 1973; Briggs, 1974). The rings partially block radio emission from Saturn's disk, thereby establishing that their optical depth at radio wavelengths is comparable to that at visual wavelengths (Schloerb et al., 1979).

The ring particles are efficient scatterers of centimeter wavelength radar signals (Goldstein and Morris, 1973). Incident polarized signals are significantly depolarized after reflection, presumably the result of multiple scattering (Goldstein et al., 1977).

Considerations of cosmic abundances, the low densities of the inner satellites of Saturn, and above all, the near infrared spectral data suggest that the ring particles are mostly water ice. If so, the low radio brightness temperature, $T_B < 0.1 \, T_E$, implies that the mean particle size is < 0.1 microwave absorption lengths in ice, or < 10 m. The high radar reflectivity sets a composition independent lower limit on the size of ~ 10 cm.

Silicates cannot be a major component of the rings because the ratio of their emissivity to reflectivity at centimeter wavelengths is much too high to match the radio and radar data.

Ice coated metallic particles of any size greater than a few centimeters would satisfy the observational constraints. Metallic particles would be well outside their Roche limit, however, and probably would have collected into a few satellites.

Uranus' Rings

The rings of Uranus were discovered during an occultation observation in March 1977 (Elliot et al., 1977; Millis et al., 1977). This and subsequent occultations provide almost all of our information about these rings.

At least nine narrow rings encircle the planet. The inner eight rings are several kilometers in radial width, whereas the outermost ring has variable width (20 to 100 km). The positions of all of the occultation crossings for each ring are well fit by a precessing Keplerian ellipse. The ellipticities thus obtained range from $< 10^{-4}$ to 8×10^{-3}, the largest value belonging to the outermost (ϵ) ring (Nicholson et al., 1978; Elliot et al., 1981). The width of the ϵ ring varies linearly with radius, a fact most naturally accounted for by an increase of eccentricity with semi-major axis.

The rings have been mapped at $\lambda = 2$ μm, where methane bands drastically reduce the brightness of the planetary disk. At this wavelength, Uranus and the rings have albedos of $\sim 10^{-4}$ and 2×10^{-2}, making the rings brighter than the planet. Individual rings are not resolved but there is an azimuthal brightness variation that is plausibly attributed to the variable width of the ϵ ring (Matthews et al., 1981).

The gravitational oblateness of Uranus is accurately determined from the ring precession rates. The optical oblateness is harder to establish but recent results from a reanalysis of the stratoscope images of Uranus and from the analysis of planetary occultation cords are in good agreement. From the gravitational and optical oblatenesses and the assumption of hydrostatic equilibrium, the planet's spin period is deduced to be ~ 15.5 hr (Franklin et al., 1980; Elliot et al., 1981). This result is a significant advance because the modern spectroscopic determinations of the period range from 12 to 25 hr.

Ring Dynamics

Collisions

The collision time t_c of the ring particles is related to the optical depth and the orbital frequency by $\Omega\, t_c \sim \tau$ and is typically several hours.

The local velocity dispersion of the ring particles is determined by the condition that the rate of conversion of orbital energy into random motions balances the rate at which the energy of random motions is dissipated into heat by inelastic collisions. Inelasticity undoubtedly increases with increasing impact velocity. For $\tau < 1$, detailed calculations show that the velocity dispersion is set by the condition that approximately 50 percent of the center of mass energy be dissipated in a typical collision (Goldreich and Tremaine, 1978a).

Collisions between particles in planetary rings seem likely to be fairly inelastic even at very low impact velocities. Thus, the velocity dispersion may be close to its minimum allowable value, $V \sim \Omega s$ where s is the typical particle radius (Brahic, 1977; Goldreich and Tremaine, 1978a).

The ring thickness $Z \sim V/\Omega$. For $V \sim \Omega s$, the ring would be not much thicker than a particle monolayer.

Collisions tend to cause an increase in the radial width of rings. In the absence of external confining torques, the spreading time, t_D, is equal to the time it takes particles to random walk a distance equal to the ring width Δr. Thus

$$t_D \sim t_c \left(\frac{\Omega \Delta r}{V}\right)^2 . \tag{1}$$

Resonance Torques

Satellites play a dominant role in determining the morphology of a planetary ring. They exert torques on the ring material at the locations of their low-order resonances. For the simplest case, that of a circular ring and a circular satellite orbit, the strongest torques occur where the ratio of the satellite orbit period to the ring particle period equals $q/(q \pm 1)$ where q is an integer. The torque is of order

$$T_q \sim \pm \ q^2 \ \frac{G^2 \, m_s^2 \, \sigma}{\Omega^2 \, r^2} \tag{2}$$

(Goldreich and Tremaine, 1978b). Here r and σ are the radius and surface mass density evaluated at the resonance location and m_s is the satellite mass. The $\pm$ sign denotes that angular momentum is always transferred outward in these interactions since the total mechanical plus gravitational energy decreases.

The m_s^2 dependence of T_q reflects its tidal nature. Although the expression for T_q does not reveal any dependence on dissipation, without dissipation T_q would vanish. Particle collisions, which damp the satellite-induced velocity perturbations, are the ultimate source of the dissipation. Under a wide range of conditions, which includes those pertaining to planetary rings, these perturbations are critically damped. Thus the exact nature of the dissipation does not affect T_q. (Note that the resonance torque here described does not act on isolated test particles, and is thus not responsible for the formation of the Kirkwood gaps in the asteroid belt.)

Planetary rings can support long leading and trailing spiral density waves that are controlled by a combination of the Coriolis force and the ring's self-gravity. Close to a resonance the long spiral waves have wavelengths several orders of magnitude greater than the interparticle spacing. These waves can exist only on the satellite side of the resonance. The leading and trailing waves propagate toward and away from the resonance, respectively. The satellite excites the long trailing wave at the resonance and this wave carries away all of the angular momentum (positive or negative) which the resonance torque gives to the disk. The wave damps due to nonlinear and viscous effects close to the resonance, and its angular momentum is dumped into the particles. The particles on the satellite side of the resonance move toward the resonance. If the resonance torque is sufficiently large, it overcomes the tendency of particle diffusion to smooth the ring's surface density and a gap opens on the satellite side of the resonance. It seems likely that this mechanism is responsible for producing the Cassini division in Saturn's rings (Goldreich and Tremaine, 1978b). The inner edge of this gap lies very near to the 2:1 orbital resonance with Mimas. This is by far the strongest resonance in Saturn's rings and it is associated with the widest gap.

A problem that remains for future investigation is to calculate the resonance torque once a gap has begun to open. Clearly the torque is weakened by gap formation. Ultimately we hope to be able to calculate the complete evolution of a gap.

The Cassini division is produced by a large satellite that orbits well outside Saturn's rings. The structure of the Uranian rings appears to require the existence of small, as yet undiscovered, satellites that orbit within the ring system (Goldreich and Tremaine, 1979; Dermott et al., 1979). The spacing between neighboring resonances from a nearby satellite is very small. Thus, it is useful to sum the discrete resonance torques and to define the total torque on a narrow ringlet of width $\Delta r \ll r$. To do so we combine the expression for T_q with the relation $q \sim r/x \gg 1$, where x ($r \gg x \gg \Delta r$) is the separation between the satellite and the ringlet. We obtain

$$T \sim \pm \frac{G^2 \, m_s^2 \, \sigma \, r \, \Delta r}{\Omega^2 \, x^4} . \tag{3}$$

A small isolated satellite of radius R_s in the midst of a broad ring (assumed to be a monolayer with $\tau \sim 1$) would open a gap of width Δr, where

$$\frac{\Delta r}{r} \sim \left(\frac{R_s}{r}\right)^2 \left(\frac{r}{s}\right)^{2/3} . \tag{4}$$

The equilibrium gap width is derived by equating the time required for the satellite torque to clear the gap and the time required by particle diffusion to fill it in. Equation (4) provides us with a practical definition of what might be called a satellite in a ring composed of particles covering a wide size range. We consider a satellite to be any body that is large enough to open a gap whose width exceeds its radius. For a ring composed of meter size particles, this definition would be satisfied by bodies having radii greater than a kilometer.

If a broad ring contains numerous satellites, they will herd the ring particles into narrow ringlets. The ringlets will be located where the net torque from the embedded satellites vanishes. This is the basic mechanism proposed by Goldreich and Tremaine (1979) for structuring the rings of Uranus.

REFERENCES

BERGE, G. L., AND D. O. MUHLEMAN, 1973. *Ap. J.* *185*, 373.
BOBROV, M. S., 1970. In *Surfaces and Interiors of Planets and Satellites,* A. Dollfus, ed., Academic Press, New York, 377.
BRAHIC, A., 1977. *Astron. Ap. 54,* 895.

BRIGGS, F. H., 1974. *Ap. J. 189*, 367.

CHIRIKOV, B. V., 1979. *Phys. Reports 52*, 263.

COLOMBO, G., 1965. *Nature 208*, 575.

COOK, A. A., AND F. A. FRANKLIN, 1958. *Smithonian Contrib. Astrophys. 2*, 377.

DERMOTT, S. F., T. GOLD, AND A. T. SINCLAIR, 1979. *Astron. J. 84*, 1225.

ELLIOT, J. L., E. DURHAM, AND D. MINK, 1977. *Nature 267*, 328.

ELLIOT, J. L., R. G. FRENCH, J. A. FROGEL, J. H. ELIAS, D. MINK, AND W. LILLER, 1981. *Astron. J.*, forthcoming.

FRANKLIN, F. A., C. C. AVIS, G. COLOMBO, AND I. I. SHAPIRO, 1980. *Ap. J. 236*, 1031.

GOLDREICH, P., 1965. *MNRAS 130*, 159.

GOLDREICH, P., AND S. J. PEALE, 1966. *Astron. J. 71*, 425.

GOLDREICH, P., AND S. TREMAINE, 1978a. *Icarus, 34* 227.

GOLDREICH, P., AND S. TREMAINE, 1978b. *Icarus 34*, 240.

GOLDREICH, P., AND S. TREMAINE, 1979. *Nature 277*, 97.

GOLDSTEIN, R. M., R. R. GREEN, G. H. PETTENGILL, AND D. B. CAMPBELL, 1977. *Icarus 30*, 104.

GOLDSTEIN, R. M., AND G. A. MORRIS, 1973. *Icarus 20*, 260.

MATTHEWS, K., P. D. NICHOLSON, AND G. NEUGEBAUER, 1981. Preprint.

MILLIS, R. L., L. H. WASSERMAN, AND P. BIRCH, 1977. *Nature 267*, 330.

Murphy, R. E., 1977. In *The Rings of Saturn*, F. D. Palluconi ed., NASA, Washington, D.C., 65.

NICHOLSON, P. D., S. E. PERSSON, K. MATTHEWS, P. GOLDREICH, AND G. NEUGEBAUER, 1978. *Astron. J. 83*, 1240.

PEALE, S. J., P. CASSEN, AND R. T. REYNOLDS, 1979. *Science 203*, 892.

PETTENGILL, G. H., AND R. B. DYCE, 1965. *Nature 206*, 1240.

PILCHER, C. B., C. R. CHAPMAN, L. A. LEBOFSKY, AND H. H. KIEFFER, 1970. *Science 167*, 1372.

ROY, A. E., AND M. W. OVENDEN, 1954. *MNRAS 114*, 232.

ROY, A. E., AND M. W. OVENDEN, 1955. *MNRAS 115*, 296.

SCHLOERB, F. P., D. O. MUHLEMAN, AND G. L. BERGE, 1979. *Icarus, 39*, 214.

10

COMETS PRODUCE SUBMICRON
PARTICLES IN THE SOLAR SYSTEM

*S. Fred Singer
and John E. Stanley*

In the study of the solar system attention has been focused mainly on the planets, asteroids, comets, and on meteor streams that penetrate the earth's atmosphere. Since the beginning of the space age, missions have also been devoted to plasmas in space made up of charged particles of atomic dimensions. Whereas the larger bodies are controlled by gravitational forces, plasma particles are controlled by electromagnetic forces.

There is a group of particles, intermediate in size and mass, whose existence has never been conclusively established. Particles with dimensions of the wavelength of visible light and smaller do not contribute to the zodiacal light and cannot be observed by standard astronomical methods. In spacecraft experiments, only extremely sensitive impact sensors hold out some hope for their detection.

Their existence has been uncertain also on theoretical grounds. Being so small, they become more sensitive to nongravitational forces, such as solar radiation pressure and electromagnetic forces. Simple considerations make this clear: gravitational force depends on

This essay was based on work supported in part by NASA Grants NSG=1256 and ·NSG=1427.

the mass of the particle and therefore on r^3, the cube of its radius; radiation pressure forces depend on the surface area, and therefore on r^3. Electromagnetic forces depend on the electric charge of the particle; assuming that the electric potential is independent of size, the charge is given by the potential times the electric capacitance, with the latter proportional to r, the first power of the radius. Thus, as particles become smaller, the gravitational force becomes progressively less important in relation to radiation pressure and electromagnetic forces.

An important problem relates to the origin of such hypothetical submicron-sized particles. Are they fragments of larger particles produced in collisions? Are they dust released by comets? Or could interstellar dust penetrate into the solar system and provide it with submicron particles? Another set of questions relates to the dynamics of such particles moving in the solar system under the combined influence of gravitational and nongravitational forces. How stable are their orbits? What is their "lifetime"?

A quite separate set of questions has to do with possible influences of such particles on the earth and on events pertaining to the earth. For example, Walter O. Roberts (Hale, 1977) has long held that incoming particles may provide "triggers" for weather phenomena. Fred Hoyle has speculated about cometary particles having structures and properties of viral particles and causing pandemics (Hoyle and Wickramasingh, 1977).

The MTS Experiment

Some of these questions have been answered recently by satellite observations. An experimental group at the NASA Langley Research Center headed by J. Alvarez prepared an extremely sensitive detector for the MTS (Meteoroid Technology Satellite) Explorer 46. The detector itself is a solid-state capacitor, consisting of a metal film, a silicon dioxide dielectric, and a silicon substrate (Kassel, 1973). The threshold of the detector is set by the thickness of the dielectric, which was either 0.4 micron or 1.0 micron. Differences in the impact rate observed by the two sets of detectors can then be attributed to

Table 10.1. Summary of the Small Meteoroid Detection Experiment

Explorer 46 (MTS)		
Altitude	660 km	
Inclination	38°	
Eccentricity	.02	
Detectors:		
Dielectric thicknesses	0.4 μm	1.0 μm
Threshold (at 15 km/sec)	10^{-17} gm	10^{-15} gm
Detector area	365 cm²	548 cm²
Raw impact rate	~2/day	~1.2/day

particles with masses of the order of 10^{-16} gram.* (See Table 10.1 for details of the experiment.)

In analyzing the Explorer 46 data (see Table 10.2), we observed a distinct difference in the background counting rates, with the less sensitive detector showing a lower counting rate, thus indicating conclusively the continuous presence of submicron particles in the vicinity of the earth.

During the nearly six months of detector operation (August 1974 through January 1975), enhancements in counting rates were observed during certain meteor showers, but not during others (see Table 10.2). Enhancements were generally most pronounced for submicron particles (except for the Leonids). The degree of enhancement seems to be inversely related to the time elapsed since the most recent perihelion passage of the comet associated with the meteor stream.** The results, though sparse, are statistically significant and leave no doubt that submicron particles are affiliated with at least some meteor streams.

This result is unexpected. There have been no prior reports of such

*A particle of specific gravity 2.0, weighing 10^{-15} gram, has a diameter of 10^{-5} centimeters (or 0.1 micron, or 1000Å). It contains about 10^7 atoms of atomic mass of about 20.

**It is significant that the Taurids, which are believed to be associated with Encke's comet, show this strong enhancement. Encke passes near the sun every three years. On the other hand, the Orionids associated with Halley's comet show no effect. Halley, of course, has not been near the sun for some seventy years but is expected to come back in 1985.

Table 10.2. Observed Flux of SubMicron Particles (in units of $10^{-4}m^{-2}sec^{-1}$)

Sensor dielectric thickness		0.4 μm	1.0 μm
Background rate*		4.6	2.0
Orionids (Halley)	Oct. 21	7.8	0.3
Taurids (Encke)	Nov. 4	27.3	5.7
Leonids (1866 I)	Nov. 17	10.1	7.1
Geminids (____)	Dec. 14	—	5.6
Ursids (Tuttle)	Dec. 22	10.3	2.2
Quadrantids (____)	Jan. 3	2.6	1.4

*Data taken over six days centered on predicted shower maximum except for Background Rate which was taken over six months.

associations of submicron particles with meteor streams, even though instruments of adequate sensitivity have been exposed to the space environment. Furthermore, observations of comet tails have shown no evidence for submicron particles (Hanner, 1980). Finally, theoretical expectations have been that submicron particles released from comets, usually near perihelion, would be immediately ejected from the solar system by radiation pressure (Harwit, 1963).

"Lifetime" of Submicron Particles

The fact that such particles are observed in meteor streams leads immediately to some upper limits on the effects of radiation pressure, and indirectly on the composition, density, and shape of the particles. It is customary to define the ratio of radiation pressure to solar gravity as β. Prior calculations had shown that β exceeds 1.0 as particles become smaller, except for particles that are dielectrics. For example, if the refractive index is below ~ 1.20, then even the presence of absorption cannot obtain a β-value > 1.0. In fact, β reaches a maximum for particles of radius around 0.1 μm, and then declines for smaller particle radii (Bandermann, 1967).* The particles could either be organic molecules, synthesized from free radicals in the nu-

*For the orbit of Encke's comet the critical value of β is 0.08. Evidently, the particles are transparent to optical radiation rather than opaque, and experience radiation pressure which is at most a few percent of the sun's gravitational force.

cleus of the comet, or silicate particles (with the impurity absorption centers split off, as suggested to us by Thomas Gold). They cannot have too low a density or too nonspherical a shape.

Aside from radiation pressure, electromagnetic forces are of major importance in removing the submicron particles from the orbit of the meteor stream. Since both Lorentz force and convective drag vary as the first power of the particle radius, they soon exceed radiation pressure and Poynting-Robertson effects. Convective drag, which relates to the solar wind velocity, is always at least an order of magnitude greater than the Lorentz force that relates to the particle's Keplerian velocity (Bandermann and Singer, 1967).

"Lifetime" is defined as the time required to remove a particle from the meteor stream orbit. It will depend on the detailed structure of the interplanetary magnetic field. For a sector field and nonzero average azimuthal field, and for particles of radius ~ 0.1 μm, this time is ~ 3 years, or 10^8 seconds (under certain assumptions about the particle and about interplanetary conditions, leading to a particle potential of $+ 3.3$ volts) (Bandermann, 1967; Bandermann and Singer, 1967). This calculated lifetime accords with the observed high counting rate for the Taurids (corresponding to Encke's comet passage of <1 year earlier) and with the essentially zero rate for the Orionids (corresponding to Halley's passage and presumed injection of particles sixty-five years ago).

Using these observations in support of a calculated lifetime, one can derive a budget for submicron particles, leading to an injection of the order of several thousand tons of material for a cometary perihelion passage (see Table 10.3). This value is in good accord with the injection mass derived from direct observations of comet tails (Sekanina, 1980).

Better data are urgently needed in order to draw more definitive conclusions. The ideal platform is an LDEF (Long Duration Exposure Facility) launched by a space shuttle, which would allow counting rates of several hundred per day during meteor showers. These data would make it quite feasible to study the decay of submicron particles with time (assuming they are all injected by the comet's perihelion passage), or the year-to-year variability, including any spreading of the stream due to electromagnetic and other forces.

Table 10.3. Sample Mass Injection Calculation for Encke

Observed flux (of particles $<10^{-15}$ gm)	$F \sim 3 \times 10^{-3}$ m^{-2} sec^{-1}
Nominal velocity	$V \sim 30$ km/sec
$\therefore$ Spatial concentration (near earth)	$N \sim 10^{-7}$ m^{-3}
Lifetime $\sim$ Taurids period $\sim$ 3.3 years	$T \sim 10^8$ sec
$\therefore$ Injection rate $\sim 1/NT$	$I \sim 10^{-15}$ m^{-3} sec^{-1}
Volume $\sim 10^{11} \times 10^{10} \times 10^{10}$ m^3	Vol $\sim 10^{31}$ m^3

Total mass injected $\sim 10^{-15} \times 10^{31} \times 10^{-15} \sim 10$ gm/sec
Or $\sim 10^9$ gm per orbital period
Or $\sim (2-5) \times 10^9$ gm at perihelion

Halley's Comet Experiment

Finally, this experiment affords an ideal opportunity to study Halley's comet. Baseline studies should be made well in advance of 1985 and continue for several years thereafter. Observations on the mass distributions, the temporal decay, and the spatial spreading can be combined to give unique information on the origin and nature of the very smallest particles released by comets into interplanetary space.

Present plans call for the first LDEF to be launched in early 1984, carrying, among other payloads, an MTS-type interplanetary dust experiment (IDE) for the detection of submicron particles. With IDE we should be able to measure the pre-Halley background. A later LDEF will give us the opportunity to observe the detailed buildup of particles as the comet passes near the sun in 1985–1986.

Together with optical observations of Halley's comet, a European flyby mission named GIOTTO, and a possible U.S. Halley intercept mission, we should be able to get a reasonably comprehensive view of a rare astronomical event.

REFERENCES

BANDERMANN, L. W., 1967. Ph.D. thesis, U. of Maryland Tech. Report 771.

BANDERMANN, L. W., AND S. F. SINGER, 1967. NASA SP-150.

HALE, L. C., 1977. *Nature 268*, 710.

HANNER, M. S., 1980. In *Solid Particles in the Solar System*, I. Halliday and B. A. McIntosh, eds., Reidel, Boston.

HARWIT, M. J., 1963. *Geophys. Res. 68*, 1271.

HOYLE, F., AND C. WICKRAMASINGH, *New Scientist*, Nov. 1977, 402; and Sept. 1978, 946.

KASSEL, P., 1973. NASA TN-D-7359.

SEKANINA, Z., 1980. in *Solid Particles in the Solar System*, I. Halliday and B. A. McIntosh, eds., Reidel, Boston.

11

GOLD'S HYPOTHESIS AND THE ENERGETICS OF THE JOVIAN MAGNETOSPHERE

A. J. Dessler

In 1976, Thomas Gold published a characteristically short paper ("The Magnetosphere of Jupiter," *J. Geophys. Res. 81*, 3401) in which he states, "The magnetosphere of Jupiter has its energy supply dominated by the rotation of the planet rather than by the stream velocity of the solar wind." His paper characteristically contains no references to the opinions of others, to whom he left the task of working out the details. Again, characteristically, Gold is correct, and I will briefly outline how power is extracted from the kinetic energy of the rotation of Jupiter and delivered to its magnetosphere. I should state, in candor, that the following is being received by the scientific community with the same enthusiasm that generally greets one of Gold's new ideas.

Any suggestion from Gold on a magnetospheric physics topic has to be taken seriously. He, after all, invented the word "magnetosphere"; he was the first to propose the mechanism of "flux-tube interchange" (now called magnetospheric convection); and he pointed out that it was the means for transporting plasma and solar-wind energy throughout the terrestrial magnetosphere (Gold, 1959).

A rather simple picture can be put forth for the extraction of power from the kinetic energy of rotation of Jupiter and the subsequent distribution of this power to the Jovian magnetosphere. This picture is based on the magnetic-anomaly model of the Jovian magnetosphere (see Dessler and Vasyliunas, 1979; Vasyliunas and Dessler, 1981, as well as the references contained therein).

Corotating Magnetospheric Convection

Two basic postulates of the magnetic-anomaly model are that the plasma that populates the magnetosphere is drawn from internal sources (the satellite Io is now known to be the principal source of plasma mass), and that the energy for driving various magnetospheric phenomena is drawn from the kinetic energy of rotation of Jupiter (e.g., Gold, 1976; Carbary et al., 1976). This situation contrasts with the earth's magnetosphere, for which energy and (some, if not most) plasma is supplied by the solar wind.

The magnetic-anomaly model is developed from the observation of the significantly nondipolar nature of the Jovian magnetic field (Smith et al., 1976; Acuña and Ness, 1976). From these data, an extensive magnetic anomaly in which the magnetic field at the surface is unusually weak has been identified by Dessler and Hill (1979). This anomaly is in the northern hemisphere and is centered at a Jovigraphic longitude of about λ_{III} (1965) $\sim 200°$ at surface levels and $\lambda_{III} \sim 230°$ at Io's orbit. It extends over a longitude range of about $120°$; this longitude interval is referred to as the active sector (Vasyliunas, 1975).

The magnetic anomaly, which is produced by high-order magnetic multipoles of Jupiter's internal magnetic field, can produce a gross asymmetry in the distant Jovian magnetosphere by at least the following two mechanisms:

(a) The area of a magnetic tube of flux entering the weak-field region of the magnetic anomaly has a larger cross-sectional area at ionospheric heights than at longitudes outside the anomaly region. This effect is schematically illustrated in Figure 11.1. Ionospheric plasma can escape more effectively from the anomaly region because of the relatively larger flux-tube foot (Dessler and Hill, 1975). Although this plasma contributes directly to an enhanced plasma density

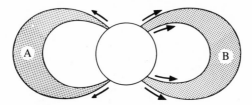

Figure 11.1. Schematic illustration of the effect of a magnetic anomaly on the flow of ionospheric plasma into the outer magnetosphere. The two flux tubes A and B are on the same L shell and have the same equatorial cross-section. However, in the longitude range of the magnetic anomaly, flux-tube B has a larger foot area than flux-tube A; the flow of ionospheric plasma from the magnetic anomaly into flux tube B is correspondingly larger than the related flow into A (after Dessler and Hill, 1975).

within the active sector, there is an important interaction that can cause the asymmetry in plasma density to be more pronounced; un-ionized gas escaping from Io is more likely to be ionized within the active sector where the extra escaping ionospheric plasma, all of which is centrifugally accelerated, provides an effective additional source of ionization. The Io plasma torus is observed to be brightest in SII emission (and hence the singly-ionized sulfur concentration is a maximum) within the longitude range of the active sector (Trafton, 1980; Pilcher and Morgan, 1980; and Trauger et al., 1980).

   (b) Within the active sector, the magnetic mirror altitude for trapped energetic particles is a minimum, and the loss cone for precipitating auroral particles is a maximum. Thus, the Jovian ionosphere is preferentially bombarded in this longitude range, which should cause the conductivity of the ionosphere and Birkeland currents to be locally enhanced (Dessler and Hill, 1979). The Birkeland current density within the active sector reaches a magnitude such that current-driven instabilities could be produced to further accelerate charged particles within the active section (e.g., Smith and Goertz, 1978; Dessler and Hill, 1979; Dessler, 1980). Such energized charged particles provide yet another ionizing source that causes gas escaping from Io to be ionized preferentially within the active sector, thus contributing to the longitudinal asymmetry in torus density described in (a).

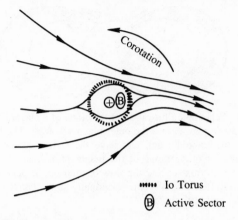

Io Torus
Active Sector

Figure 11.2. The denser portion of the Io plasma torus (the portion of the torus in the active sector) is flung outward by centrifugal force to drive a magnetospheric convection pattern that corotates with Jupiter (Vasyliunas, 1978; Hill, 1980a; and Hill et al, 1981). Shown is a sketch of the magnetospheric region out to about 25 R_J. Plasma from the region of the active sector, labeled B to correspond to flux tube B of Figure 11.1, flows away from Jupiter to drive the convection pattern. This entire pattern is fixed relative to the active sector, and hence the convection pattern corotates with Jupiter even though the plasma itself might not corotate (Hill, 1980b).

Corotating Convection and Magnetospheric Energization

The asymmetrical torus drives a magnetospheric convection pattern that corotates with Jupiter (Vasyliunas, 1978; Hill, 1980a; and Hill et al., 1981), as schematically illustrated in Figure 11.2. The Io torus is thus not only the principal source of magnetospheric plasma mass, but it is the agent for extracting kinetic energy of rotation from Jupiter to drive the broad array of Jovian magnetospheric phenomena (Dessler, 1980; Eviatar and Siscoe, 1980; and Dessler, 1981).

Because of Jupiter's rapid rotation, the Io torus has a significant potential energy available to drive magnetospheric convection. The torus is in a centrifugal potential field $\Omega^2 r$ where Ω is the angular rotation rate of Jupiter and r is the Jovicentric distance to an element of torus plasma. At $6R_J$, $\Omega^2 r = 13.1$ m/sec^2 or 1.3 g's. In falling through the centrifugal potential from $6R_J$ to L_o (the effective limit of corotation), the torus releases to magnetospheric convection an amount of energy

$$W = M_t \int_{6R_J}^{L_o} \Omega^2 \, r \, dr = M_t \, \Omega^2 \, (L_o^2 - 36R_J^2)/2, \tag{1}$$

where R_J is the radius of Jupiter.

If this potential energy is extracted in a time Δt, the power supplied to magnetospheric convection is $P_t = W/\Delta t$. Note that Δt is effectively the time to renew the torus, or, Δt can be regarded as the residence time of an ion in the torus.

From (1)

$$P_t = W/\Delta t = (M_t/\Delta t)\Omega^2(L_o^2 - 36R_J^2)/2, \tag{2}$$

from which we obtain

$$dM/dt = M_t\Delta t = 2P_t/[\Omega^2(L_o^2 - 36R_J^2)]. \tag{3}$$

Thus, if we evaluate (3), we can obtain an independent value of the mass loading rate. We know Ω and R_J quite well; L_o has been determined by Hill (1980b) to within a factor a 2; and estimates of magnetospheric power dissipation are available to yield a value for P_t.

By fitting the experimental results from Voyager 1 of Bridge et al (1979) to his theoretical model, Hill (1980b) estimates L_o to be about 20 R_J. Then, to obtain an estimate for P_t, we neglect unobservable loss processes such as Joule heating of the ionosphere caused by convection related ionospheric currents and, as a lower limit, set P_t equal to the energy of the observed phenomena, thus enabling us to obtain a lower limit for dM/dt.

The most energetic of the observed magnetospheric phenomena appears to be the Jovian aurora that is associated with the Io torus (Sandel et al., 1979). Sandel and his colleagues assume the aurora is caused by energetic electrons having a total power input of 1.7×10^{14} W. (They have recently reduced this power estimate to 1.3×10^{13} W [Sandel, private communication].) Goertz (1980), has pointed out that if the bombarding particles are protons, the energy input is less certain. He derives a range of auroral power inputs of between 6×10^{11} W to 10^{14} W, with 6×10^{12} W as his best estimate. Not many of the magnetospheric ions are protons, however (sulfur and oxygen predominate), and it is not clear how this fact affects Goertz's estimate.

We provisionally adopt a value $P_t = 3 \times 10^{13}$ W, which allows for other, presently observed and unobserved loss processes. Substituting $P_t = 3 \times 10^{13}$ W, $\Omega = 1.76 \times 10^{-4}$ rad/sec, $L_o = 20$, and $R_J = 7.14 \times 10^7$ m into (3), we obtain $dM/dt = 1.0 \times 10^3$ kg/sec, in agreement with other estimates of the mass loading rate. Thus, if Io injects approximately 1 ton/sec into the Io torus, enough energy will be extracted from the kinetic energy of rotation of Jupiter to power all observed phenomena plus a few presently unobserved phenomena.

Discussion

The two most striking effects of the corotating magnetospheric convection are (a) the release of relativistic electrons from the Jovian magnetosphere into interplanetary space exhibiting a modulation period equal to Jupiter's spin period (Chenette et al., 1974) and (b) a persistent bulge of atomic hydrogen in Jupiter's upper atmosphere which is fixed in System III coordinates and therefore corotates with the planetary magnetic field (Dessler et al., 1981). There are now nearly a dozen Jovian phenomena that have been related, in one way or another, to the magnetic-anomaly model. It appears that, as pointed out by Gold (1976), the Jovian magnetosphere is "intermediate between the Earth's magnetosphere and that of pulsars." Jupiter's magnetosphere is rather pulsar-like in that both plasma and energy are drawn from internal sources and not the solar wind, a situation quite different from the earth's magnetosphere, where energy and (most?) plasma is supplied by the solar wind. We can thus expect to learn from Jupiter some physics that will help us gain a better insight into the physics of pulsars.

REFERENCES

DESSLER, A. J., 1980. *Icarus 44*, 291.
DESSLER, A. J., AND T. W. HILL, 1975. *Geophys. Res. Lett. 2*, 567.
DESSLER, A. J., AND T. W. HILL, 1979. *Astrophys. J. 227*, 664.

DESSLER, A. J., AND V. M. VASYLIUNAS, 1979. *Geophys. Res. Lett. 6*, 37.

DESSLER, A. J., B. R. SANDEL, AND S. K. ATREYA, 1981. *Planet. Space Sci. 29*, 215.

EVIATAR, A., AND G. L. SISCOE, 1980. *Geophys. Res. Lett. 7*, 1085.

GOERTZ, C. K., 1980. *J. Geophys. Res. 85*, 2949.

GOERTZ, C. K., *Geophys. Res. Lett. 7*, 365.

GOLD, T., 1959. *J. Geophys. Res. 64*, 1219.

GOLD, T., 1976. *J. Geophys. Res. 81*, 3401.

HILL, T. W., 1980a. In "Conference on the Physics of the Jovian Magnetosphere," Rice Univ., Houston,

HILL, T. W., 1980b. *Science 207*, 301.

HILL, T. W., 1981. *EOS, Transactions AGU 62*, 25.

HILL, T. W., A. J. DESSLER, AND L. J. MAHER, 1981. *J. Geophys. Res. 86*, 9020.

PILCHER, C. B., AND J. S. MORGAN, 1979. *Science 205*, 297.

SANDEL, B. R., D. E. SHEMANSKY, A. L. BROADFOOT, J. L. BERTAUX, J. E. BLAMONT, M. J. S. BELTON, J. B. HOLBERG, S. K. ATREYA, T. M. DONAHUE, H. W. MOOS, D. F. STROBEL, J. C. MCCONNELL, A. DALGARNO, R. GOODY, AND M. B. MCELROY, 1979. *Science 206*, 962.

SMITH, E. J., L. DAVIS, JR., AND D. E. JONES, 1976. *Jupiter*, T. Gehrels, ed., Univ. of Arizona Press, Tucson, 788.

SMITH, ROBERT A., AND C. K. GOERTZ, 1978. *J. Geophys. Res. 83*, 2617.

TRAFTON, L., 1980. *Icarus 42*, 111.

TRAUGER, J. T., G. MUNCH, F. L. ROESLER, 1980. *Astrophys. J. 236*, 1035.

VASYLIUNAS, V. M., 1978, In *Cospar Program/Abstracts*, Innsbruck, Austria, 29 May–10 June, 66.

VASYLIUNAS, V. M., 1975. *Geophys. Res. Lett. 2*, 87.

VASYLIUNAS, V. M., AND A. J. DESSLER, 1981. *J. Geophys. Res. 86*, 8435.

Plate 4. The Coma cluster of galaxies. Courtesy, Kitt Peak National Observatory.

Appendix

BIBLIOGRAPHY OF SCIENTIFIC
PUBLICATIONS BY THOMAS GOLD

1947

1. Transient reception and the degree of resonance of the human ear. R. J. Pumphrey and T. Gold. *Nature 160*, 124.

1948

2. Hearing. I. The cochlea as a frequency analyzer. With R. J. Pumphrey. *Proc. Roy. Soc. B 135*, 462–491.
3. Hearing. II. The physical basis of the action of the cochlea. *Proc. Roy. Soc. B 135*, 492–498.
4. Phase memory of the ear: a proof of the resonance hypothesis. R. J. Pumphrey and T. Gold. *Nature 161*, 640.
5. The steady-state theory of the expanding universe. H. Bondi and T. Gold. *Mon. Not. Roy. Astron. Soc. 108*, 252–270.

1949

6. Creation of matter in the universe. *Nature 164*, 1006.
7. Rotation and terrestrial magnetism. *Nature 163*, 513.

The editors thank Steven Soter for preparing this bibliography.

8. A thermodynamic consideration in relation to acoustic energy in stellar models. *Mon. Not. Roy. Astron. Soc. 109*, 115.

1950

9. A design of an ultrasonic delay-line. *Phil. Mag 42*, 787.

1951

10. Dynamo action. *Proceedings of the Conference on Dynamics of Ionized Media* (Dept. of Physics, University College, London).
11. The origin of cosmic radio noise. *Proceedings of the Conference on Dynamics of Ionized Media* (Dept. of Physics, University College, London).

1952

12. The alignment of galactic dust. *Mon. Not. Roy. Astron. Soc. 112*, 215.
13. The "double bang" of supersonic aircraft. *Nature 170*, 808.
14. Is there an aether? H. Bondi and T. Gold. *Nature 169*, 146.
15. Polarization of starlight. *Nature 169*, 322.
16. Relativity and indeterminancy. H. Bondi and T. Gold. *Nature 170*, 582.

1953

17. Relation between modern cosmologies and nuclear astrophysical processes. *Mem. Soc. Roy. Sci. Liége, 4th Ser. 13*, 68.

1954

18. A note on the reported color-index effect of distant galaxies. H. Bondi, T. Gold, and D. W. Sciama. *Astrophys. J. 120*, 597.
19. Suggestions for rocket astronomy. *Rocket Exploration of the Upper Atmosphere* (Pergamon Press, London), pp. 366–367.

20. Theories of instellar polarization. *Mem. Soc. Roy. Sci. Liége, 4th Ser.* *15*, 591.

21. Concluding remarks on turbulence in the interstellar gas. *Gas Dynamics of Cosmic Clouds* (I.A.U. Symposium No. 2; H. C. van de Hulst and J. M. Burgers, eds.; North Holland Publ., Amsterdam), pp. 238–240.

22. Discussion on shock waves and rarified gases. *Gas Dynamics of Cosmic Clouds* (I.A.U. Symposium No. 2; H. C. van de Hulst and J. M. Burgers, eds.; North Holland Publ., Amsterdam), pp. 103–104.

1955

23. The field of a uniformly accelerated charge, with special reference to the problem of gravitational acceleration. H. Bondi and T. Gold. *Proc. Roy. Soc. A 229*, 416.

24. The "horizon" of the steady-state universe. *Nature 175*, 382.

25. Instability of the Earth's axis of rotation. *Nature 175*, 526.

26. The lunar surface. *Mon. Not. Roy. Astron. Soc. 115*, 585–604.

27. On the damping of the free nutation of the Earth. H. Bondi and T. Gold. *Mon. Not. Roy. Astron. Soc. 115*, 41.

28. The symmetry of the corona of 1954 June 30. *Mon. Not. Roy. Astron. Soc. 115*, 340.

1956

29. Cosmology. *Vistas in Astronomy 2* (A. Beer, ed.; Pergamon Press, London and New York), pp. 1722–1726.

30. High-energy particles in solar flares. *Nature 178*, 487.

31. Ojämnheter i jordens rotation. *Särtryck ur Populär Astronomisk Tidskrift*, Parts 1–2, 20.

32. The solar outburst, 23 February 1956. Observations by the Royal Greenwich Observatory. With D. R. Palmer. *J. Atmosph. Terrestr. Phys. 8*, 287

1957

33. On the gravitational interaction of matter and antimatter. P. Morrison and T. Gold. *Babson Award Essays*. Unpublished.

1958

34. The arrow of time. *Structure and Evolution of the Universe* (Proc. Solvay Conf.; R. Stoops, ed.; Institut International de Physique Solvay), pp. 1–15.
35. The cosmic ray flare effect. G. Cocconi, K. Greisen, P. Morrison, T. Gold, and S. Hayakawa. *Nuovo Cimento 8*, 161.
36. Cosmic rays from the Sun. *Smithson. Inst. Ann. Rept. 1957*, 233.
37. Irregularities in the Earth's rotation. *Sky & Tel. 17* (Nos. 5 and 6, March and April).
38. The magnetic field in the corona. *Electromagnetic Phenomena in Cosmical Physics* (I.A.U. Symposium No. 6; B. Lehnert, ed.; Cambridge University Press, Cambridge), pp. 275–279.
39. Origin of tektites. *Nature 181*, 172–74.

1959

40. Cosmic rays and radio waves as manifestations of a hot universe. With F. Hoyle. *Paris Symposium on Radio Astronomy* (I.A.U. Symposium No. 9; R. N. Bracewell, ed.; Stanford Univ. Press, Stanford), pp. 583–588.
41. Dust on the Moon. *Vistas in Astronautics 2* (Pergamon Press, London), pp. 261–266.
42. Magnetic field in the solar system. *Nuovo Cimento 13*, 318–323.
43. Motions in the magnetosphere of the Earth. *J. Geophys. Res. 64*, 1219–1224.
44. Origin of the radiation near the Earth discovered by means of satellites. *Nature 183*, 355–358.
45. Plasma and magnetic fields in the solar system. *J. Geophys. Res. 64*, 1665–1674.
46. Space research in relation to the Moon and the nearer planets. *Proc. Roy. Soc. A 253*, 487–491.

1960

47. Abundance of lithium and origin of the solar system. *Astrophys. J. 132*, 274–275.
48. Cosmic garbage. *Air Force and Space Digest 43*(5), 65.

49. Energetic particle fluxes in the solar system and near the Earth. *Astrophys. J. Suppl. 4*, 406–416.

50. On the origin of solar flares. With F. Hoyle. *Mon. Not. Roy. Astron. Soc. 120*, 89–105.

51. On the origin of the solar system. *From Nucleus to Universe* (S. T. Butler and H. Messel, eds.; Shakespeare Head Press, Sydney), pp. 189–217.

52. Solar whislters. With D. H. Menzel. *Radio Noise Spectrum* (D. H. Menzel, ed.; Harvard Univ. Press, Cambridge), pp. 123–128.

53. Interstellar hydrogen. *Radio Noise Spectrum* (D. H. Menzel, ed.; Harvard Univ. Press, Cambridge), pp. 177–179.

1961

54. The Moon. *Space Astrophysics* (W. Liller, ed.; McGraw Hill, New York), pp. 172–178.

55. Present evidence concerning magnetic fields and particle fluxes in the solar system. T. Gold. *Space Research II* (H. C. van de Hulst, C. de Jager, and A. F. Moore, eds.; North Holland Publ., Amsterdam), pp. 828–836.

56. The problem of the abundance of the hydrogen molecule. *Mem. Soc. Roy. Sci. Liége 20*, 476–481.

57. Solar power in space. *Astronautics*, Feb. 1961, 34, 35, 68.

1962

58. The arrow of time. *Amer. J. Phys. 30*, 403–410.

59. Cosmic rays and the interplanetary medium. *Astronautics*, Aug. 1962, 43–45.

60. Emission from the Sun, and transmission of radiation through the interplanetary space. *J. Phys. Soc. Japan 17*, Suppl. A-II, 607–609.

61. Fields and particles in interplanetary space. *Bolletino Soc. Ital. Fisica 24*, 181–193.

62. Interchange and rotation of the Earth field lines. *J. Phys. Soc. Japan 17*, Suppl. A-I, 187–189.

63. Magnetic storms. *Space Sci. Rev. 1*, 100–114.

64. Processes on the lunar surface. *The Moon* (I.A.U. Symposium No. 14;

Z. Kopal and Z. K. Mikhailov, eds.; Academic Press, London and New York), pp. 433–439.

65. The propagation of solar particles to the Earth. *J. Phys. Soc. Japan 17*, Suppl. A-II, pp. 600–605.

1963

66. The first five years of space research. *The Universe of Space and Time* (S. T. Butler and H. Messel, eds.; Shakespeare Head Press, Sydney), pp. 263–291.
67. The interstellar abundance of the hydrogen molecule. II. Galactic abundance and distribution. R. J. Gould, T. Gold, and E. E. Salpeter. *Astrophys. J. 138*, 408–425.
68. Large solar outbursts in the past. *Pontificiae Academiae Scientiarum Scripta Varia 25*, 159–165.
69. Magnetic field configurations near the Sun and in interplanetary space. *Pontificiae Academiae Scientiarum Scripta Varia 25*, 431–457.
70. Problems requiring solution. *Origin of the Solar System* (R. Jastrow and A. G. W. Cameron, eds.; Academic Press, New York and London), pp. 171–174.

1964

71. A new joint American-Australian astronomy center. With H. Messel. *Nature 204*, 18–20.
72. Magnetic energy shedding in the solar atmosphere. *AAS-NASA Symposium on the Physics of Solar Flares* (W. N. Hess, ed.; NASA SP-50), pp. 389–395.
73. Outgassing processes on the Moon and Venus. *The Origin and Evolution of Atmospheres and Oceans* (P. J. Brancazio and A. G. W. Cameron, eds.; John Wiley, New York), pp. 249–256.
74. Ranger Moon pictures: implications. *Science 145*, 1046–1048.
75. The structure of the lunar surface: a survey of the present evidence. *Lectures in Aerospace Medicine* (U.S.A.F. School of Aerospace Medicine, Brooks A.F.B., Texas), pp. 240–271.
76. Structure of the Moon's surface. *The Lunar Surface Layer* (J. S. Salisbury and P. E. Glaser, eds.; Academic Press, New York and London), pp. 345–353.

1965

77. The arrow of time. *Time* (S. T. Butler and H. Messel, eds.; Shakespeare Head Press, Sydney), pp. 143–165.

78. The dissipation of magnetic energy in the upper atmosphere of the Sun. *Stellar and Solar Magnetic Fields* (R. Lust, ed.; North Holland Publ., Amsterdam), pp. 390–397.

79. The expansion of a relativistic gas into intergalactic space. *Proceedings of the Ninth International Conference on Cosmic Rays* (Inst. of Phys. and Physical Soc. Publ., London), pp. 132–134.

80. Rotation of the planet Mercury. S. J. Peale and T. Gold, *Nature 206,* 1240–1241.

81. The stability of multistar systems. With W. I. Axford and E. C. Ray. *Quasi-stellar Sources and Gravitational Collapse* (Robinson et al., eds.; Univ. Chicago Press, Chicago), pp. 93–98.

82. Cosmic processes and the nature of time. *Mind and Cosmos: Essays in Contemporary Science and Philosophy* (R. G. Colodny, ed.; Univ. Pittsburgh Press, Pittsburgh), pp. 311–329.

1966

83. More reactions to Surveyer-1: first impressions of the implications. *New Scientist,* 16 June 1966, 699.

84. Long-term stability of the Earth-Moon system. *The Earth-Moon System* (B. G. Marsden and A. G. W. Cameron, eds.; Plenum Press, New York), pp. 93–97.

85. Luna 9 pictures: implications. With B. W. Hapke. *Science 153,* 290–293

86. The magnetosphere of the Moon. *The Solar Wind* (R. J. Mackin, Jr., and M. Neugebauer, eds.; Jet Propulsion Laboratory, Pasadena), pp. 381–391.

87. Maser action in interstellar OH. F. Perkins, T. Gold, and E. E. Salpeter. *Astrophys. J. 145,* 361–366.

88. The Moon's surface. *The Nature of the Lunar Surface* (W. N. Hess, D. H. Menzel, and J. A. O'Keefe, eds.; The Johns Hopkins Univ. Press, Baltimore), pp. 107–121.

89. Probable mode of landing of Luna 9. *Nature 210,* 150–151.

90. The purpose of space research. *Proceedings of the Fifth National Conference on the Peaceful Uses of Space* (NASA SP-82), pp. 63–68.

1967

91. The evolution of dense star systems. *High Energy Astrophysics* (Les Houches Symposium, Gordon and Breach, London), pp. 239–241.
92. *The Nature of Time* (T. Gold, ed.; Cornell Univ. Press, Ithaca).
93. On the detection of water on the Moon. M. Werner, T. Gold, and M. Harwit. *Planet. Space Sci. 15*, 771–774.
94. Quasi-stellar objects and radio galaxies. *Astrophys. J. 147*, 833.
95. Radio method for the precise measurement of the rotation of the Earth. *Science 157*, 302–304.

1968

96. Can the observed microwave background be due to a superposition of sources? With F. Pacini. *Astrophys. J. 152*, L115–L118.
97. Density of the lunar surface. M. J. Campbell, J. Ulrichs, and T. Gold. *Science 159*, 973.
98. General consequences of the magnetic field dissipation theory of flares. *Mass Motions in Solar Flares and Related Phenomena* (Y. Öhman, ed.; John Wiley, New York), pp. 205–210, 153.
99. Goals of space: science, prestige, or economic advantage. *The Quality of Life* (Cornell Univ. Press, Ithaca), pp. 98–104.
100. Jodrell Bank: one man's vision is the world's gain. *Scientific Research 16*(3), 48–49.
101. Lens effect of the Earth's ionosphere on radio waves reaching the Moon. With G. Silvestro. *Planet. Space Sci. 16*, 999–1009.
102. Lunar theory and processes. D. E. Gault, R. J. Collins, T. Gold, J. Green, G. P. Kuiper, H. Masursky, J. O'Keefe, R. Phinney, and E. M. Shoemaker. *J. Geophys. Res. 73*, 4115–4131.
103. Maser action in space. *Interstellar Ionized Hydrogen* (Y. Terzian, ed.; W. A. Benjamin, New York), pp. 747–761.
104. Rotating neutron stars as the origin of the pulsating radio sources. *Nature 218*, 731–732.
105. A search for lunar outgassing. With M. Simon, and J. J. Condon. *Planet. Space Sci. 16*, 825.
106. Surveyer V: chemical observations. *Science 160*, 904–905.

1969

107. Apollo 11 observations of a remarkable glazing phenomenon on the lunar surface. *Science 165*, 1345–1349.
108. Atmospheric tides and the resonant rotation of Venus. With S. Soter. *Icarus 11*, 356–366.
109. The nature of pulsars: survey of present views. *Proceedings of the International Symposium on Contemporary Physics* (International Atomic Energy Agency, Vienna), pp. 477–481.
110. The new planetary astronomy. *Technology Review 72*, 42–45.
111. The pulsar enigma. *Pulsating stars 2* (Macmillan, London), pp. x–xii.
112. Pulsars and the mass spectrum of cosmic rays. *Nature 223*, 162.
113. Rotating neutron stars and the nature of pulsars. *Nature 221*, 25–27.

1970

114. Apollo 11 and 12 close-up photography. *Icarus 12*, 360–375.
115. Apollo 12 seismic signal: indication of a deep layer of powder. With S. Soter. *Science 169*, 1071–1075.
116. Lunar surface closeup steroscopic photography. With F. Pearce and R. Jones. *Apollo 12 Preliminary Science Report* (NASA SP-235), pp. 183–188.
117. Lunar theory and processes: chemical observations by Surveyor V. *Icarus 12*, 224–225.
118. Lunar theory and processes: the physical condition of the lunar surface. *Icarus 12*, 226–229.
119. Optical and high-frequency electrical properties of the lunar sample. With M. J. Campbell and B. T. O'Leary. *Science 167*, 707–709.
120. Origin of glass deposits in lunar craters. *Science 168*, 611.
121. Pulsars and the origin of high energy particles. *Proceedings of the 11th International Conference on Cosmic Rays* (G. Bozoki et al., eds.; Central Research Institute for Physics, Budapest), pp. 163–175.
122. Statistical investigation of the distribution of pulsars in space. With H. M. Newman. *Nature 227*, 151–152.

1971

123. Atmospheric tides and the 4-day circulation on Venus. With S. Soter. *Icarus 14*, 16–20.

124. Evolution of mare surface. *Proc. 2nd Lunar Sci. Conf.*, 2675–2680.

125. Lunar-surface closeup stereoscopic photography. *Apollo 14 Preliminary Science Report* (NASA SP-272), pp. 239–247.

126. Magnetically trapped particles in the lower solar atmosphere. A. O. Benz and T. Gold. *Solar Physics 21*, 157–166.

127. The nature of the surface of the moon. *Space Res. 11*, 51–61.

128. The nature of the lunar surface: recent evidence. *Proc. Amer. Phil. Soc. 115*, 74–82.

129. Pulsars and the origin of cosmic rays. *Highlights of Astronomy* (C. de Jager, ed.; D. Reidel Publ., Dordrecht-Holland), pp. 727–730.

130. Some physical properties of Apollo 12 lunar samples. With B. T. O'Leary and M. Campbell. *Proc. 2nd Lunar Sci. Conf.*, 2173–2181.

1972

131. Erosion, transportation and the nature of the maria. *The Moon* (I.A.U. Symposium No. 47; H. Urey and K. Runcorn, eds.; D. Reidel Publ., Dordrecht-Holland), pp. 55–67.

132. Grain size analysis, optical reflectivity measurements, and determination of high-frequency electrical properties for Apollo 14 lunar samples. With E. Bilson and M. Yerbury. *Proc. 3rd Lunar Sci. Conf.*, 3187–3193.

133. Introduction to the Viking Symposium. *Icarus 16*, ii–iii.

134. The origin of pulsar radiation. *Proceedings of the Symposium on Pulsars and High Energy Activity in Supernovae Remnants* (Accademia Nazionale dei Lincei, Rome), pp. 161–166.

135. Rotating neutron stars and the nature of pulsars. *The Physics of Pulsars* (A. M. Lenchek, ed.; Gordon and Breach, New York), pp. 101–109.

1973

136. Conjectures about the evolution of the Moon. *The Moon 7*, 293–306.

137. Electrostatic transportation of dust on the Moon. With G. J. Williams.

Photon and Particle Interactions with Surfaces in Space (R. J. L. Grard, ed.; D. Reidel Publ., Dordrect-Holland), pp. 557–560.

138. Grain size analysis and high frequency electrical properties of Apollo 15 and 16 samples. With E. Bilson and M. Yerbury. *Proc. 4th Lunar Sci. Conf.*, 3093–3100.

139. Multiple universes. *Nature 242*, 24–25.

140. Particle interactions with celestial objects—concluding remarks. *Photon and Particle Interactions with Surfaces in Space* (R. J. L. Grard, ed.; D. Reidel Publ., Dordrecht-Holland), pp. 571–576.

141. The simulation of lunar micrometeorite impacts by laser pulses. E. Bilson, T. Gold, and G. Gull. *The Moon 6*, 405–413.

142. Sputtering and darkening of the grains on the lunar surface. *Photon and Particle Interactions with Surfaces in Space* (R. J. L. Grard, ed.; D. Reidel Publ., Dordrecht-Holland), pp. 517–519.

1974

143. Extrasolar planetary systems. *Communication with Extraterrestrial Intelligence (CETI)* (C. Sagan, ed.; MIT Press, Cambridge), pp. 8–23.

144. The Moon. *New Science in the Solar System* (New Science Publications, London), pp. 25–36.

145. The movement of small particulate matter in the early solar system and the formation of satellites. *Highlights of Astronomy*, vol. 3 (G. Contopoulos, ed.; D. Reidel Publ., Dordrecht-Holland), pp. 483–485.

146. Observation of iron-rich coating on lunar grains and a relation to low albedo. With E. Bilson and R. L. Baron. *Proc. 5th Lunar Conf. 3*, 2413–2422.

147. On the exposure history of the lunar regolith. With G. J. Williams. *Proc. 5th Lunar Conf. 3*, 2387–2395.

148. Optical properties of the Apollo 15 deep core samples. With E. Bilson and R. L. Baron. *Proc. 5th Lunar Conf. 3*, 2355–2359.

149. Pulsars and the origin of cosmic rays. *Phil. Trans. Roy. Soc. London A 277*, 453–471.

150. Skylab: Is it worth the risk and the expense? *Bull. Atomic Scientists 30*, 4–8.

151. The world map and the apparent flow of time. *Modern Developments in Thermodynamics* (B. Gal-Or, ed.; Israel Universities Press, Jerusalem), pp. 63–72.

1975

152. After dinner talk: How *not* to do science. T. Gold. *Ann. New York Acad. Sci. 262*, 496–500.

153. Auger analysis of the lunar soil: study of processes which change the surface chemistry and albedo. With E. Bilson and R. L. Baron. *Proc. 6th Lunar Sci. Conf.*, 3285–3303.

154. The origin of the terrestrial planets. *Our Earth* (H. Messel and S. T. Butler, eds.; Shakespeare Head Press, Sydney), pp. 247–257.

155. The Earth and the Moon. *Our Earth* (H. Messel and S. T. Butler, eds.; Shakespeare Head Press, Sydney), pp. 258–277.

156. The planets inside the Earth's orbit: the Earth's sister Venus, and the small, dense planet Mercury. *Our Earth* (H. Messel and S. T. Butler, eds.; Shakespeare Head Press, Sydney), pp. 278–289.

157. The most tantalising of the planets: Mars. *Our Earth* (H. Messel and S. T. Butler, eds.; Shakespeare Head Press, Sydney), pp. 290–302.

158. Mother and baby paradox. *Nature 256*, 113.

159. Remarks on the paper "The tidal loss of satellite-orbiting objects and its implications for the lunar surface" by Mark J. Reid. *Icarus 24*, 134–135.

160. Resonant orbits of grains and the formation of satellites. *Icarus 25*, 489–491.

1976

161. Cometary impact and the magnetization of the Moon. With S. Soter. *Planet. Space Sci. 24*, 45–54.

162. Electrical properties of Apollo 17 rock and soil samples and a summary of the electrical properties of lunar material at 450 MHz frequency. With E. Bilson and R. L. Baron. *Proc. 7th Lunar Sci. Conf.*, 2593–2603.

163. The magnetosphere of Jupiter. *J. Geophys. Res. 81*, 3401–3402.

164. The role of rotation in high-energy astrophysics. *High Energy Activity During the Late Phases of Stellar Evolution* (Accademia dei Lincei, Rome), pp. 95–102.

165. The surface chemical composition of lunar samples and its significance for optical properties. With E. Bilson and R. L. Baron. *Proc. 7th Lunar Sci Conf.*, 901–911.

1977

166. Chemical and optical properties at the Apollo 15 and 16 sites. With E. Bilson, R. L. Baron, M. Z. Ali, and W. D. Ehmann. *Proc. 8th Lunar Sci. Conf.*, 3633–3643.

167. Electrical properties at 450 MHz of Apollo 15 and 16 deep drill core samples and surface soil samples at the same site. With E. Bilson and R. L. Baron. *Proc. 8th Lunar Sci. Conf.*, 1271–1275.

168. On the kinetics of solar wind acceleration. A. O. Benz and T. Gold. *Astron. Astrophys. 55*, 229–237.

169. Origin and evolution of the lunar surface: the major questions remaining. *Phil. Trans. Roy. Soc. Lond. A 285*, 555–559.

170. The relationship of surface chemistry and albedo of lunar soil samples. With E. Bilson and R. L. Baron. *Phil. Trans. Roy. Soc. Lond. A 285*, 427–431.

171. Relativity and time. *The Encyclopaedia of Ignorance* (R. Duncan and M. Weston-Smith, eds.; Pergamon Press, Oxford), p. 100.

172. The rings of Uranus: theory. S. F. Dermott and T. Gold. *Nature 267*, 590–593.

173. The search for the cause of the low albedo of the Moon. With E. Bilson and R. L. Baron. *J. Geophys. Res. 82*, 4899–4908.

174. The surface composition of lunar soil grains: a comparison of the results of Auger and X-ray photoelectron (ESCA) spectroscopy. R. L. Baron, E. Bilson, T. Gold, R. J. Colton, B. Hapke, and M. A. Steggert. *Earth Planet. Sci. Lett. 37*, 263–272.

1978

175. On the origin of the Oort cloud. S. F. Dermott and T. Gold. *Astron. J. 83*, 449–450.

176. Response to "Comments on the surface composition of lunar soil grains" by R. M. Housley. R. L. Baron, E. Bilson, T. Gold, R. J. Colton, B. Hapke, and M. A. Steggert. *Earth Planet. Sci. Lett. 41*, 471–472.

1979

177. Arbitrarily slow irreversibility: note on Lynden-Bell's example. *Observatory 99*, 45–46.

178. Brontides: natural explosive noises. With S. Soter. *Science 204*, 371–375.

179. Determination of secondary electron emission characteristics of lunar soil samples. With R. L. Baron and E. Bilson. *Earth Planet. Sci. Lett. 45*, 133–140.

180. The earthquake evidence for Earth gas. *Energy for Survival* (H. Messel, ed.; Pergamon Press Australia), pp. 65–78.

181. The supply of natural fuels. *Energy for Survival* (H. Messel, ed.; Pergamon Press Australia), pp. 79–92.

182. Electrical origin of the outbursts on Io. *Science 206*, 1071–1073.

183. North Seaquakes. S. Soter and T. Gold. *New Scientist 83*, 542.

184. The rings of Uranus: nature and origin. S. F. Dermott, T. Gold, and A. T. Sinclair. *Astron. J. 84*, 1225–1234.

185. Terrestrial sources of carbon and earthquake outgassing. *J. Petrol. Geol. 1*(3), 3–19.

186. Theory of the Earth-synchronous rotation of Venus. With S. Soter. *Nature 277*, 280–281.

1980

187. The deep-Earth-gas hypothesis. With S. Soter. *Scientific American 242*(6), 154–160.

188. The deep Earth gas hypothesis. With S. Soter. *Bull. Verein. Schweiz. Petroleum-Geologen und -Ingenieure 46*, 11–35.

189. Outgassing and the power source for plate tectonics. *Mechanisms of Continental Drift and Plate Tectonics* (P. A. Davies and S. K. Runcorn, eds.; Academic Press, London), pp. 343–344.

1981

190. Natural explosive noises. With S. Soter. *Science 212*, 1297–1298.

1982

191. Abiogenic methane and the origin of petroleum. With S. Soter. *Energy Exploration & Exploitation*, forthcoming.

CONTRIBUTORS

SIR HERMANN BONDI, FRS, is Professor of Mathematics at King's College, University of London, and Chairman of the Natural Environment Research Council, London. He is also Past Director General of the European Space Research Organization and Chief Adviser in the Ministry of Defense of the United Kingdom.

ALEXANDER J. DESSLER, Ph.D. 1955, Duke University, is currently Professor and Chairman, Department of Space Physics and Astronomy, Rice University. His primary research interest is magnetospheric physics, with recent emphasis on the magnetosphere of Jupiter. In June 1982 he assumed duties as Director of the Space Science Laboratory at NASA Marshall Space Flight Center in Huntsville, Alabama.

PETER GOLDREICH is currently Professor of Planetary Science and Astronomy at the California Institute of Technology, Pasadena. He does research and teaching in theoretical astrophysics and geophysics. He is a disciple of Thomas Gold, who introduced him to many of the mysteries of the universe.

DAVID J. HELFAND is a member of the faculty of the Department of Astronomy at Columbia University. He is a specialist on radio and X-ray emission from pulsars.

SIR FRED HOYLE, FRS, is Honorary Professor of Physics and Astronomy, Manchester University, and Past Director of the Institute of Theoretical Astronomy, Cambridge University. He was Andrew D. White Professor-at-Large, Cornell University, from 1972 to 1978.

EDWIN E. SALPETER is James Gilbert White Distinguished Professor in Physical Sciences and Director of the Center for Radiophysics and Space Research, Cornell University. He is a member of the U.S. National Science Board.

DENNIS SCIAMA is a Senior Research Fellow of All Souls College at Oxford. He is a specialist in cosmology and relativistic astrophysics.

S. FRED SINGER is Professor of Environmental Sciences, University of Virginia. His research interests include energy resources and the interplanetary dust. He is now working on the Interplanetary Dust Experiment to be flown on the NASA long duration exposure space station in 1984.

JOHN E. STANLEY is Research Associate in the Department of Environmental Sciences, University of Virginia. His main research concerns the interplanetary dust. He is preparing the Interplanetary Dust Experiment to be flown on the NASA long duration exposure space station in 1984.

JOSEPH H. TAYLOR was educated at Haverford College and Harvard University, obtaining his Ph.D. in astronomy in 1968. He is now Professor of Physics at Princeton University, and does research in radio astronomy, especially involving observations of pulsars.

SAUL TEUKOLSKY is a faculty member of the Departments of Physics and Astronomy at Cornell University. His research is concerned primarily with general relativity and relativistic astrophysics.

Index

Index

Index

COSMOLOGY AND
ASTROPHYSICS

Designed by G. T. Whipple, Jr.
Composed by The Composing Room of Michigan, Inc.
in 11 point Times Roman, 2 points leaded,
with display lines in Times Roman.
Printed offset by Thomson-Shore, Inc.
on Warren's Number 66 text, 50 pound basis.
Bound by John Dekker & Sons, Inc.
in Holliston book cloth
and stamped in Kurz-Hastings foil.

Library of Congress Cataloging in Publication Data
Main entry under title:

Cosmology and astrophysics.

Papers from a symposium held Oct. 9–11, 1980,
at Cornell University, Ithaca, N.Y.
"Bibliography of scientific publications by
Thomas Gold": p.
Includes index.
1. Cosmology—Addresses, essays, lectures.
2. Nuclear astrophysics—Addresses, essays,
lectures. 3. Solar system—Addresses, essays,
lectures. 4. Gold, Thomas. 5. Astronomers—
United States—Biography—Addresses, essays,
lectures. I. Gold, Thomas. II. Terzian,
Yervant, 1939–   III. Bilson, Elizabeth M.
QB985.C67   1982   523.1   82-7268
ISBN 0-8014-1497-0   AACR2